Endogenous Peptides and Centrally Acting Drugs

Progress in Biochemical Pharmacology

Vol. 16

Series Editor
R. Paoletti, Milan

S. Karger · Basel · München · Paris · London · New York · Sydney

24th Annual OHOLO Biological Conference on Neuroactive Compounds and Their Cell Receptors, Zichron Ya'acov, April 1–4, 1979

Endogenous Peptides and Centrally Acting Drugs

Volume Editors
A. Levy and E. Heldman, Ness-Ziona; *Z. Vogel,* Rehovot and
Y. Gutman, Jerusalem

Technical Editor
S.R. Smith, Ness-Ziona

52 figures and 25 tables, 1980

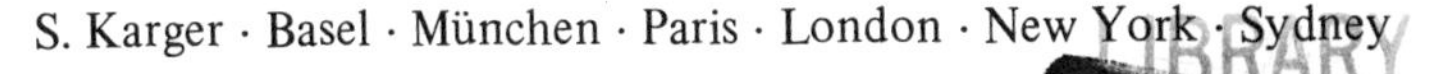

S. Karger · Basel · München · Paris · London · New York · Sydney

Progress in Biochemical Pharmacology

Vol. 14: Ecological Perspectives on Carcinogens and Cancer Control. Selected Papers of the International Conference, Cremona 1976. Eds.: Stock, C.C. (New York, N.Y.); Santamaria, L. (Pavia); Mariani, P.L. (Cremona), and Gorini, S. (Milan).
VI + 170 p., 51 fig., 42 tab., 1978 ISBN 3–8055–2684–9
Vol. 15: Lipoprotein Metabolism. Ed.: Eisenberg, S. (Jerusalem).
VI + 262 p., 46 fig., 18 tab., 1979. ISBN 3–8055–2985–6

National Library of Medicine, Cataloging in Publication
OHOLO Biological Conference, 24th, Zichron Ya'acov, Israel 1979
Endogenous peptides and centrally acting drugs
Volume editors, A. Levy [et al.]; technical editor, S.R. Smith.–
Basel, New York, Karger, 1980.
(Progress in biochemical pharmacology; vol. 16)
Consists of proceedings of the 2nd of the 24th annual OHOLO Biological Conference.
1. Peptides – metabolism – congresses 2. Neurochemistry – congresses 3. Central Nervous System Agents – metabolism – congresses I. Levy André, 1923– ed.
II. Title III. Series
Wl PR666H v.16/QU 68 038 1979e
ISBN 3–8055–0831–X

ISBN 3–8055–0831–X

Contents

Endogenous Peptides and Centrally Acting Drugs

Papers presented at the 24th Annual OHOLO Biological Conference, Zichron Ya'acov, Israel 1979, on the subject 'Neurobiology of Cholinergic and Adrenergic Transmitters' are published in Monographs in Neural Sciences, Vol. 7. For table of contents see page VI of this volume.

Neurobiology of Cholinergic and Adrenergic Transmitters

24th Annual OHOLO Biological Conference, Zichron Ya'acov, Israel 1979, published as Vol. 7 of 'Monographs in Neural Sciences'.

Conference Organizing Committee

Permanent Committee

Prof. Simon Gitter, Tel-Aviv University Sackler School of Medicine, Ramat-Aviv
Prof. Natan Grossowicz, Hebrew University, Jerusalem
Prof. Israel Hertman, Israel Institute for Biological Research, Ness-Ziona
Prof. Alexander Keynan, Hebrew University, Jerusalem
Prof. Marcus A. Klingberg, Israel Institute for Biological Research, Ness-Ziona and Tel-Aviv University Sackler School of Medicine
Prof. Alexander Kohn, Israel Institute for Biological Research, Ness-Ziona and Tel-Aviv University Sackler School of Medicine
Prof. Michael Sela, Weizmann Institute of Science, Rehovot

Organizing Committee (24th Meeting)

Prof. Y. Gutman, Hebrew University – Hadassah Medical School, Jerusalem
Dr. Y. Grunfeld, Israel Institute for Biological Research, Ness-Ziona
Dr. E. Heldman, Israel Institute for Biological Research, Ness-Ziona
Prof. Israel Hertman, Israel Institute for Biological Research, Ness-Ziona
Prof. Marcus A. Klingberg, Israel Institute for Biological Research, Ness-Ziona and Tel-Aviv University Sackler School of Medicine
Prof. Alexander Kohn, Israel Institute for Biological Research, Ness-Ziona and Tel-Aviv University Sackler School of Medicine
Dr. A. Levy, Israel Institute for Biological Research, Ness-Ziona
Dr. Zvi Paster, Israel Institute for Biological Research, Ness-Ziona
Dr. Z. Vogel, Weizmann Institute of Science, Rehovot

Ms. Ariela Snir, Secretary

Mr. Mordechai Navon, Technical Organizer

Acknowledgments

The Conference was organized and sponsored by the *Israel Institute for Biological Research,* Ness-Ziona, Israel
affiliated to the Tel-Aviv University
Sackler School of Medicine

The financial support given to the conference by the following institutions is gratefully acknowledged:

Ames Division Miles Laboratory Inc. (USA)
'Avi' Foundation, Tel-Aviv University (Israel)
Bank Hapoalim Ltd., Tel-Avi (Israel)
The British Council (Israel and UK)
Ciba-Geigy, Basel (Switzerland)
Israel Center for Psychobiology, Charles G. Smith Family Foundation (Israel)
Dr. Karl Thomas GmbH (BRD)
Upjohn Company (USA)

List of Participants

Assael, M., Kaplan Hospital, Rehovot (Israel)
Abraham, S., Israel Institute for Biological Research, Ness-Ziona (Israel)
Abramsky, O., Hadassah University Hospital, Jerusalem (Israel)
Altstein, Miriam, The Weizmann Institute of Science, Rehovot (Israel)
Amir, Adina, Israel Institute for Biological Research, Ness-Ziona (Israel)
Amiran, Y., National Council for Research and Development (Israel)
Amitai, G., Israel Institute for Biological Research, Ness-Ziona (Israel)
Angel., I., Tel-Aviv University, Ramat Aviv (Israel)
Anglister, Lili, The Weizmann Institute of Science, Rehovot (Israel)
Ashkenazi, Ruth, Technion – Israel Institute of Technology Haifa (Israel)
Avissar Sofia, Tel-Aviv University, Ramat Aviv (Israel)
Aylon, D., Ichilov Hospital, Tel-Aviv (Israel)
Bahar, Leah, The Weizmann Institute of Science, Rehovot (Israel)
Barr-Nea, Liliane, Tel-Aviv University, Ramat Aviv (Israel)
Bauminger, Sara, Ichilov Hospital, Tel-Aviv (Israel)
Bechar, M., Beilinson Hospital, Petah Tikva (Israel)
Belmaker, R., Ezrath Nashim Hospital, Jerusalem (Israel)
Ben-Barak, Y., The Weizmann Institute of Science, Rehovot (Israel)
Ben-David, A., Medical Corps, Israel Defence Forces (Israel)
Ben-David, Nita, Israel Institute for Biological Research, Ness-Ziona (Israel)
Benhar, E., The Weizmann Institute of Science, Rehovot (Israel)
Ben-Ishay, D., Ikapharm-Plantex, Kfar Sava (Israel)
Ben-Zvi, Z., The Ben Gurion University of the Negev, Beersheva (Israel)
Berger, B., Haifa University, Haifa (Israel)
Bigalke, H., University of Giessen, Giessen (FRG)
Blum, Ilana, Beilinson Hospital, Petah Tikva (Israel)
Blumberg, S., The Weizmann Institute of Science, Rehovot (Israel)
Bracha, H., Ezrath Nashim Hospital, Jerusalem (Israel)
Brenner, Talma, Hadassah University Hospital, Jerusalem (Israel)
Bruckstein, Rachel, Sackler School of Medicine, Tel-Aviv University, Ramat Aviv (Israel)
Burgen, Sir A., National Institute for Medical Research, London (England)
Burnstock, G., University College London, London (England)
Coq, H., D.R.E.T., Paris (France)
Cordova, Tova, Ichilov Hospital, Tel-Aviv (Israel)

Danon, A., The Ben Gurion University of the Negev, Beersheva (Israel)
De Baetselier, Annie, The Weizmann Institute of Science, Rehovot (Israel)
De Vries, A., Tel-Aviv University, Ramat Aviv (Israel)
Dodiuk, Hanna, Israel Institute for Biological Research, Ness-Ziona (Israel)
Dudai, Y., The Weizmann Institute of Science, Rehovot (Israel)
Duvdevani Nurit, Israel Institute for Biological Research, Ness-Ziona (Israel)
Ebstein, R., Ezrath Nashim Hospital, Jerusalem (Israel)
Edery, H., Israel Institute for Biological Research, Ness-Ziona, and Sackler School of Medicine, Tel-Aviv (Israel)
Egozi, Y., Tel-Aviv University, Ramat Aviv (Israel)
Elizur, A., Shalvata Hospital, Hod Hasharon (Israel)
Evenchik, Z., Israel Institute for Biological Research, Ness-Ziona (Israel)
Feinberg, J., Technion – Israel Institute of Technology, Haifa (Israel)
Fink, G., The Weizmann Institute of Science, Rehovot (Israel)
Fisher, A., Israel Institute for Biological Research, Ness-Ziona (Israel)
Friedman, J., The Weizmann Institute of Science, Rehovot (Israel)
Fuchs, P., Israel Institute for Biological Research, Ness-Ziona (Israel)
Fuchs, Sara, The Weizmann Institute of Science, Rehovot (Israel)
Gainer, H., National Institutes of Health, Bethesda, Md. (USA)
Galron, Ronit, Tel-Aviv University, Ramat Aviv (Israel)
Gazit, H., The Weizmann Institute of Science, Rehovot (Israel)
Ginzberg, Irit, The Weizmann Institute of Science, Rehovot (Israel)
Gitter, S., Tel-Aviv University, Ramat Aviv (Israel)
Gaathon, A., Ames Yissum Ltd., Jerusalem (Israel)
Goldberg, Ora, The Weizmann Institute of Science, Rehovot (Israel)
Granoth, I., Israel Institute for Biological Research, Ness-Ziona (Israel)
Green, I., Medical Corps, Israel Defence Forces (Israel)
Grunfeld, Y., Israel Institute for Biological Research, Ness-Ziona (Israel)
Gurwitz, D., Tel-Aviv University, Ramat Aviv (Israel)
Gutman, H., Medical Corps, Israel Defence Forces (Israel)
Gutman, Y., The Hebrew University – Hadassah Medical School Jerusalem (Israel)
Haber, R., Abic Ltd., Ramat Gan (Israel)
Habermann, E., University of Giessen, Giessen (FRG)
Habermann, F., Israel Institute for Biological Research, Ness-Ziona (Israel)
Halperin, Bilha, Israel Institute for Biological Research, Ness-Ziona (Israel)
Hamburger, A.D., Rafa Laboratories, Jerusalem (Israel)
Hamon, M., Collège de France, Paris (France)
Hanani, M., Hadassah University Hospital, Jerusalem (Israel)
Heldman, E., Israel Institute for Biological Research, Ness-Ziona (Israel)
Heron, D., The Weizmann Institute of Science, Rehovot (Israel)
Hertman, I., Israel Institute for Biological Research, Ness-Ziona (Israel)
Hermoni, Myra, Ezrath Nashim Hospital, Jerusalem (Israel)
Herz, A., Max-Planck-Institut für Psychiatrie, Munich (FRG)
Hoffman, Lynne, The Weizmann Institute of Science, Rehovot (Israel)
Hovevy-Zion, D., Israel Institute for Biological Research, Ness-Ziona (Israel)
Inbar, A., Eldan Electronic Instruments, Jerusalem (Israel)
Ishaaya, I., Volcani Center, Bet Dagon (Israel)
Israeli, E., Israel Institute for Biological Research, Ness-Ziona (Israel)
Jacobowitz, D., National Institutes of Mental Health, Bethesda, Md. (USA)
Johnson, M.K., M.R.C. Toxicology Unit, Carshalton, Surrey (England)

Kalir, A., Israel Institute for Biological Research, Ness-Ziona (Israel)
Kalmus, Y., Blumenthal Hospital, Haifa (Israel)
Kaplanski, J., The Ben Gurion University of the Negev, Beersheva (Israel)
Kaplun, A., Israel Institute for Biological Research, Ness-Ziona (Israel)
Katz, D., Rafa Laboratories, Jerusalem (Israel)
Katz, D., Israel Institute for Biological Research, Ness-Ziona (Israel)
Katz, Esther, Israel Institute for Biological Research, Ness-Ziona (Israel)
Katz, P., Israel Institute for Biological Research, Ness-Ziona (Israel)
Keenan, A., University College, Dublin (Ireland)
Keren, Ora, Tel-Aviv University, Ramat Aviv (Israel)
Keysary, A., Israel Institute for Biological Research, Ness-Ziona (Israel)
Kiezman, G., Abic Ltd., Ramat Gan (Israel)
Klingberg, M. A., Institute for Biological Research, Ness-Ziona (Israel)
Klug, Y., Tel-Aviv University, Ramat Aviv (Israel)
Koch, Y., The Weizmann Institute of Science, Rehovot (Israel)
Kohn, Chana, Kupat Holim, Rehovot (Israel)
Kohn, A., Israel Institute for Biological Research, Ness-Ziona (Israel)
Kosterlitz, H.W., University of Aberdeen, Aberdeen (Scotland)
Kullmann, R., University of Bonn, Bonn (FRG)
Kushnir, M., Israel Institute for Biological Research, Ness-Ziona (Israel)
Lachman, C., Israel Institute for Biological Research, Ness-Ziona (Isarel)
Lampert, A., The Weizmann Institute of Science, Rehovot (Israel)
Lawson, T., Radiochemical Centre, Amersham (England)
Leader, H., Israel Institute for Biological Research, Ness-Ziona (Israel)
Leibar, Sara, Beilinson Hospital, Petah Tikva (Israel)
Lerner, Sarah, The Hebrew University – Hadassah Medical School, Jerusalem (Israel)
Levy, A., Israel Institute for Biological Research, Ness-Ziona (Israel)
Levy, Miriam, Tel-Aviv University, Ramat Aviv (Israel)
Liron, Z., Israel Institute for Biological Research, Ness-Ziona (Israel)
Littauer, U., The Weizmann Institute of Science, Rehovot (Israel)
Lotan, Ilana, Tel-Aviv University, Ramat Aviv (Israel)
Loyter, A., The Hebrew University of Jerusalem, Jerusalem (Israel)
Luini, A., The Weizmann Institute of Science, Rehovot (Israel)
Magnes, J., The Hebrew University of Jerusalem, Jerusalem (Israel)
Mandel, P., Centre National de la Recherche Scientifique, Strasbourg (France)
Manulis, Shulamit, Volcani Center, Bet Dagon (Israel)
Meidan, Rina, The Weizmann Institute of Sicnece, Rehovot (Israel)
Michaelson, D., Tel-Aviv University, Ramat Aviv (Israel)
Miller, E., Department of Health, Education and Welfare, Washington, D.C. (USA)
Modai, I., Gehah 'Pat' Hospital, Petah Tikva (Israel)
Monzain, Rachel, Israel Institute for Biological Research, Ness-Ziona (Israel)
Nadler, Etta, Tel-Aviv University, Sackler School of Medicine, Ramat Aviv (Israel)
Navon, M., Israel Institute for Biological Research, Ness-Ziona (Israel)
Neubaur, I., Volcani Center, Bet Dagon (Israel)
Ophir, Idith, Tel-Aviv University, Ramat Aviv (Israel)
Oppenheim, Bina, Technion – Israel Institute of Technology Haifa (Israel)
Paster, Z., Israel Institute for Biological Research, Ness-Ziona (Israel)
Patchornik, A., The Weizmann Institute of Science, Rehovot (Israel)
Peled, M., Israel Institute for Biological Research, Ness-Ziona (Israel)
Pinchasi, Irit, Tel-Aviv University, Ramat Aviv (Israel)

Pollard, H.B., National Institutes of Health, Bethesda, Md. (USA)
Porath, Gila, Israel Institute for Biological Research, Ness-Ziona (Israel)
Rabey, J., Ichilov Hospital, Tel-Aviv (Israel)
Rahamimoff, R., Hebrew University – Hadassah Medical School, Jerusalem (Israel)
Ravid, Roma, Ichilov Hospital, Tel-Aviv (Israel)
Rehavi, M., Tel-Aviv University, Ramat Aviv (Israel)
Rekhes, A., Hadassah University Hospital, Jerusalem (Israel)
Rosedky-Gratz, Yael, Beilinson Hospital, Petah Tikva (Israel)
Samuel, D., The Weizmann Institute of Science, Rehovot (Israel)
Santenac, Ms., D.R.E.T., Paris (France)
Scherson, Talma, The Weizmann Institute of Science, Rehovot (Israel)
Schidrut, Ester, Hadassah Hospital, Tel-Aviv (Israel)
Schneeweiss, F., The Weizmann Institute of Science, Rehovot (Israel)
Schoene, K., Institut für Aerobiologie, Schmallenberg-Grafschaft (FRG)
Schou, M., Aarhus University, Risskov (Denmark)
Schuldiner, S., Hebrew University – Hadassah Medical School, Jerusalem (Israel)
Sedvall, G., Karolinska Institute, Stockholm (Sweden)
Shahar, A., Israel Institute for Biological Research, Ness-Ziona (Israel)
Shainberg, A., Bar-Ilan University, Ramat Gan (Israel)
Sherman-Gold, Rivka, The Weizmann Institute of Science, Rehovot (Israel)
Shchory, Mina, Hebrew University – Hadassah School of Medicine, Jerusalem (Israel)
Shod, Freda, Abic Ltd., Ramat Gan (Israel)
Shofer, H., Israel Institute for Biological Research, Ness-Ziona (Israel)
Shtager, Z., Medical Corps, Israel Defence Forces (Israel)
Silman, I., The Weizmann Institute of Science, Rehovot (Israel)
Simon, G., Israel Institute for Biological Research, Ness-Ziona (Israel)
Simantov, R., The Weizmann Institute of Science, Rehovot (Israel)
Sinai, Judith, Israel Institute for Biological Research, Ness-Ziona (Israel)
Sivan, Dvora, Tel-Aviv University, Ramat Aviv (Israel)
Snir, Ariela, Israel Institute for Biological Research, Ness-Ziona (Israel)
Smith, Shirley R., Israel Institute for Biological Research, Ness-Ziona (Israel)
Sokolovsky, M., Tel-Aviv University, Ramat Aviv (Israel)
Sporn, E., U.S. Food and Drug Administration, Washington, D.C. (USA)
Steinberg, I.Z., The Weizmann Institute of Science, Rehovot (Israel)
Stotzky, A., Ikapharm-Plantex, Kfar Sava (Israel)
Straussman, Yoheved, Israel Institute for Biological Research, Ness-Ziona (Israel)
Streifler, M., Tel-Aviv University, Sackler School of Medicine, Ichilov Hospital, Tel-Aviv (Israel)
Suzin, Y., Israel Institute for Biological Research, Ness-Ziona (Israel)
Szechtman, H., The Weizmann Institute of Science, Rehovot (Israel)
Teichberg, V., The Weizmann Institute of Science, Rehovot (Israel)
Teomi, Shoshana, Israel Institute for Biological Research, Ness-Ziona (Israel)
Torten, M., Israel Institute for Biological Research, Ness-Ziona (Israel)
Vardi, J., Ichilov Hospital, Tel-Aviv (Israel)
Vincze, A., Israel Institute for Biological Research, Ness-Ziona (Israel)
Vogel, Z., The Weizmann Institute of Science, Rehovot (Israel)
Wassermann, Ita, The Weizmann Institute of Science, Rehovot (Israel)
Weiner, N., University of Colorado, Denver, Colo. (USA)
Weinstock-Rozin, Marta, Hadassah Medical School, Jerusalem (Israel)
Weissman, B.A., Israel Institute for Biological Research, Ness-Ziona (Israel)
White, M., Israel Institute for Biological Research, Ness-Ziona (Israel)

Wilchek, M., The Weizmann Institute of Science, Rehovot (Israel)
Winkler, D., Barzilai Medical Center, Ashkelon (Israel)
Ya'ari, Y., Medical Corps, Israel Defence Forces (Israel)
Yavetz, Bella, Tel-Aviv University, Ramat Aviv (Israel)
Yavin, E., The Weizmann Institute of Science, Rehovot (Israel)
Yavin, Ziva, The Weizmann Institute of Science, Rehovot (Israel)
Youdim, M., Technion, Faculty of Medicine, Technion – Israel Institute of Technology, Haifa (Israel)
Zahavy, J., Israel Institute for Biological Research, Ness-Ziona (Israel)
Zinder, O., Rambam Medical Center, Haifa (Israel)
Zisapel, Nava, Tel-Aviv University, Ramat Aviv (Israel)
Zor, U., The Weizmann Institute of Science, Rehovot (Israel)
Zurgil, Neomi, Tel-Aviv University, Ramat Aviv (Israel)
Zutra, Aliza, The Weizmann Institute of Science, Rehovot (Israel)

OHOLO Conferences

1956 Bacterial Genetics
1957 Tissue Cultures in Virological Research
1958 Inborn and Acquired Resistance to Infection in Animals
1959 Experimental Approach to Mental Diseases
1960 Cryptobiotic Stages in Biological Systems
1961 Virus-Cell Relationships
1962 Biological Synthesis and Function of Nucleic Acids
1963 Cellular Control Mechanism of Macromolecular Synthesis
1964 Molecular Aspects of Immunology
1965 Cell Surfaces
1966 Chemistry and Biology of Psychotropic Agents
1967 Structure and Mode of Action of Enzymes
1968 Growth and Differentiation of Cells *in vitro*
1969 Behaviour of Animal Cells in Culture
1970 Microbial Toxins
1971 Interaction of Chemical Agents with Cholinergic Mechanism
1972 New Concepts in Immunity in Viral and Rickettsial Diseases
1973 Strategies for the Control of Gene Expression
1974 Sensory Physiology and Behavior
1975 Air Pollution and the Lung
1976 Host-Parasite Relationships in Systemic Mycoses
1977 Skin: Drug Application and Evaluation of Environmental Hazards
1978 Extrachromosomal Inheritance in Bacteria
1979 Neuroactive Compounds and Their Cell Receptors

Introductory Address

Ladies and Gentlemen,

It is a pleasure and a privilege to open the 24th Annual OHOLO Conference on 'Neuroactive Compounds and their Cell Receptors'. We extend a very warm welcome to the participants and guests from abroad and from Israel.

The OHOLO Conferences were initiated 24 years ago by the Israel Institute for Biological Research, and take their name from the site of the first meetings on the shores of Lake Kinneret. The purpose of these meetings is, as it was then, 'to foster interdisciplinary communication between scientists in Israel, and to provide added stimulus by the participation of invited scientists from abroad'. Topics of particular interest to scientists involved in basic and applied research are chosen. We try to provide a relaxed atmosphere with ample time for informal as well as formal discussions.

Each year there is a steady growth in the number of Conference participants, due to the international reputation it has gained. The Conference is organized as a joint effort by the regular staff of our Institute. For us OHOLO has become synonymous with spring time, and every spring brings new developments in our ever-changing reality.

In my opening address at last year's Conference, which was conducted with the sound of cannons and airplanes in the background, I stressed that even though we had come together in a period of violence, our ongoing efforts to achieve scientific and technological progress for the benefit of our people and the peoples in the area could not be abandoned or delayed without endangering our fundamental hope for peace and security. This spring we perceive with hope the first fragile blossom of peace.

Let me briefly introduce our Institute to this audience. We are a small community of scientists and technologists whose aim is to strengthen the link in our country between advanced academic research and its application in industry and environmental health. Our Department of Pharmacology, which organized this Conference, is a pioneer in this endeavor. It initiated and is now enjoying fruitful cooperation with the Israeli chemical and pharmaceutical industries, and is in the process of establishing an independent international Toxicology Testing Laboratory. Our Division of Chemistry recently initiated an advanced Chemical Analytical Center involved in pharmacokinetics, detection of environmental pollutants and complex chemical analyses. Next year's OHOLO Conference will deal with 'New Developments in Human and Veterinary Vaccines', a topic of interest to our Biology Division. Together with our activities in applied research we maintain high scientific standards, and our scientists are welcomed in institutions of international reputation.

The 24th OHOLO Conference will review recent advances in Neurobiology, and deal with recent research on various aspects of neuroactive compounds. The multidisciplinary character of neurobiology is well expressed by the physiologists, biochemists, pharmacologists and morphologists gathered here to present their own points of view and their current research on many of the aspects with which Neurobiology is concerned today. The topics to be discussed in the sessions are: understanding of neurotransmitter release; properties of various agonists and antagonists and how these properties affect the binding of ligands to their receptors; recent developments in psychopharmacology; and finally, considerable attention will be given to the rapidly developing area of neuroactive peptides.

The invited papers have been chosen to present the various topics from a fairly broad perspective and are intended to be of interest to research workers in related fields as well as to those involved in clinical studies.

We look forward to discussing these important and exciting topics during the course of the Conference, and we sincerely hope that it will contribute to progress in Neurobiology and its clinical application.

We gladly take this opportunity to express our thanks to Dr. *E. Heldman*, Ms. *Ariella Snir* and the Organizing Committee for their efforts and dedication in organizing the meeting. We express our gratitude for the invaluable assistance of Mr. *M. Navon* as technical organizer and to the supplementary staff who contributed to the success of the Conference.

I. Hertman
Israel Institute for Biological Research
Ness-Ziona

Prog. biochem. Pharmacol., vol. 16, pp. 3–10 (Karger, Basel 1980)

Opioid Peptides and Their Receptors[1]

H. W. Kosterlitz

Unit for Research on Addictive Drugs, University of Aberdeen, Aberdeen

Introduction

It is now generally accepted that there are two separate and independent systems of opioid peptides in the central and peripheral nervous systems. The long-chain peptide, β-endorphin, is present in the system that is centred around the hypothalamus-pituitary axis and extends in a well-defined pathway into the midline regions of the diencephalon and anterior pons (3, 17). In contrast, the short-chain peptides, methionine- and leucine-enkephalin, are widely but unevenly spread throughout the brain, spinal cord and peripheral autonomic nervous system (4, 6, 7, 15). In such circumstances, the question arises of whether or not the two types of opioid peptides interact with one and the same receptor or whether there are several receptors as is the arrangement in the catecholamine system. Already before the discovery of the endogenous opioid peptides, *Martin* and his colleagues (12, 13), from experiments on the chronic spinal dog, had adduced evidence for the view that the action of compounds with dual agonist and antagonist action, e.g. nalorphine and cyclazocine, and of certain benzomorphans without antagonist component, e.g. ethylketazocine, cannot be explained on the basis of a single opiate receptor and suggested the presence of μ-, κ- and σ-receptors. This view was supported by investigations on the *in vitro* models of the guinea-pig ileum and mouse vas deferens (8); the results led to similar conclusions that, apart from the μ-receptors mediating the action of classical morphine-like compounds, receptors of at least one other type (κ-receptors) were present in the central and peripheral nervous systems.

[1] Supported by grants from the Medical Research Council, the U.S. National Institute on Drug Abuse (DA 00662) and the U.S. Committee on Problems of Drug Dependence.

Evidence for Multiple Receptors

When the problem of multiple receptors arises, several experimental approaches can be used. The first method is that of multiple parallel assays, which is indirect but can give reliable information and has been used by *Lord et al.* (11). The second depends on the specificity of antagonists, a method which has been so useful for characterization of the α- and β-adrenoceptors. The design of *Lord et al.* (11) consisted of four parallel assays, the guinea-pig ileum and mouse vas deferens as pharmacological models and the inhibition of [^{3}H]-naloxone or [^{3}H]-naltrexone binding and [^{3}H]-leucine-enkephalin binding in brain homogenates. The results of these parallel assays led to the conclusion that in the brain there are at least two types of receptors for the opioid peptides, namely the μ-receptors which represent the preferential binding sites for naloxone, naltrexone or morphine, and the δ-receptors which represent the preferential binding sites for the enkephalins. Of the two pharmacological models, the guinea-pig ileum has mainly μ-receptors and the mouse vas deferens mainly δ-receptors. This interpretation was strongly supported by the fact that, in the mouse vas deferens, the opioid peptides are difficult to antagonise by naloxone in that they require a concentration about 10 times higher than that needed for morphine. In the guinea-pig ileum the enkephalins are much less potent than in the mouse vas deferens but they do not require more naloxone for antagonism than morphine. This observation also indicated that the enkephalins can interact with the μ-receptors for which they have a lower affinity than for the δ-receptors.

Another approach is the determination of the maximum number of binding sites for a given ligand (5). There are apparently binding sites of only one affinity for D-Ala^2-D-Leu^5-enkephalin in guinea-pig brain homogenate. At this site, the ligand has a K_D of 1.27 ± 0.16 nM (n = 5) and the number of binding sites corresponds to 7.4 ± 0.31 pmol/g wet wt (n = 5). Under the same conditions, only binding sites of single affinities were detected for D-Ala^2-L-Leu^5-enkephalin amide or D-Ala^2-L-Met^5-enkephalin amide. The corresponding K_D values were 2.52 ± 0.48 and 1.96 ± 0.17 nM (n = 5) and the numbers of binding sites were 12.4 ± 0.93 and 12.8 ± 2.22 pmol/g wet wt (n = 5). Etorphine was found to have a similar number of apparently homogeneous binding sites (15.4 ± 2.4 pmol/g wet wt (n = 3)) but a higher affinity (0.37 ± 0.03 nM (n = 3)) than the D-Ala^2-enkephalin amides. However, with both morphine and dihydromorphine binding sites of high and low affinities were found; the total number of binding sites were 3.7 ± 0.59 and 4.3 ± 0.36 pmol/g wet wt (n = 4), respectively.

Yet another approach is the comparison of the maximal number of binding sites occupied by a mixture of two primary ligands with the maximal number of binding sites occupied by each ligand. When the binding of D-Ala^2-D-Leu^5-enkephalin, dihydromorphine and a mixture of the two ligands in a ratio of 3:1 was compared in the same homogenate, the mixture yielded a larger number of cpm than the maximal number of cpm produced by each of the two ligands separately; in the experimental conditions of our laboratory the difference was 1568 ± 608 cpm ($n = 5$; $p < 0.05$, paired analysis).

It is an interesting finding that the tritiated primary ligands which we have investigated can be divided into two groups. In the first group, the maximal number of binding sites varies between 4 and 7 pmol/g tissue whereas in the second there are more binding sites, i.e. 12 and 15 pmol/g tissue. In the first group, morphine interacts preferentially with μ-receptors represented by the guinea-pig ileum and has a higher affinity to the [^{3}H]-naltrexone than to the [^{3}H]-leucine-enkephalin binding sites, whereas D-Ala^2-D-Leu^5-enkephalin interacts preferentially with δ-receptors represented by the mouse vas deferens and has a higher affinity to the [^{3}H]-leucine-enkephalin binding sites.

On the other hand, the amides of D-Ala^2-L-Met^5-enkephalin and D-Ala^2-L-Leu^5-enkephalin belong to the second group, as does etorphine. It is possible that these compounds bind to both sites but the experiments designed to test this view are not yet complete.

Fragments of β-Endorphin

As far as the long-chain peptides are concerned, β-endorphin has been shown to be equipotent in depressing the contractions of the guinea-pig ileum and those of the mouse vas deferens, and is also equiactive in inhibiting the binding of the three ligands [^{3}H]-naltrexone, [^{3}H]-dihydromorphine and [^{3}H]-leucine-enkephalin in homogenates of guinea-pig brain (9).

When fragments of β-endorphin were assayed in a similar manner, it was found that in the guinea-pig ileum and for the inhibition of [^{3}H]-naloxone and [^{3}H]-dihydromorphine binding in homogenates of guinea-pig brain, β-endorphin shows the highest potency, followed by methionine-enkephalin, LPH_{61-87}, LPH_{61-79} and LPH_{61-77}, in this order. In the mouse vas deferens and for the inhibition of [^{3}H]-leucine-enkephalin binding, methionine-enkephalin is more potent than any of the other fragments, including β-endorphin (LPH_{61-91}). It may be of physiological significance that the long-chain peptide LPH_{61-91} has similar potencies in different assay systems whereas the short-chain peptide is more differentiated as far as its affinity to different receptors is concerned (9).

Activity Pattern of Enkephalin Analogues

Since the biological half-time of the two naturally occurring enkephalins is very short, many attempts have been made to design stable analogues with strong antinociceptive activity. It is therefore important to know which alterations in the molecule are permissible without concomitant changes in the pattern of pharmacological activity (10). The replacement of Gly^2 by D-Ala in leucine-

Table I. Relative potencies of analogues of opioid peptides, referred to the potencies of methionine-enkephalin to inhibit the contractions of the mouse vas deferens and the binding of [^{3}H] -leucine-enkephalin in brain homogenates

Compound	Inhibition of contractions of			Inhibition of binding in brain homogenates of		
	Guinea-pig ileum	Mouse vas deferens	Gpi/ Mvd	[^{3}H] -Nal-trexone	[^{3}H] -Leucine-enkephalin	Naltr/ Leu-enk
Tyr-Gly-Gly-Phe-L-Leu	0.043	1.47	0.029	0.047	0.76	0.06
Tyr-D-Ala-Gly-Phe-L-Leu	0.61	9.9	0.061	0.058	0.56	0.10
Tyr-D-Ala-Gly-Phe-D-Leu	0.37	30	0.012	0.038	0.34	0.11
Tyr-D-Met-Gly-Phe-Pro amide	1.14	1.67	0.68	0.22	0.12	1.8
Tyr-D-*n*Leu-Gly-Phe-Pro ethylamide	1.19	2.21	0.54	0.17	0.073	2.3
Tyr-D-Ala-Gly-NCH_3 Phe-Met(O)-ol	2.00	1.08	1.9	0.40	0.029	14
Morphine	0.19	0.027	7.0	0.066	0.008	8.3
Methionine-enkephalin	0.11	1	0.11	0.18	1	0.18

The values are the means obtained from the relative potencies (methionine-enkephalin = 1) by multiplying those for the guinea-pig ileum by the conversion factor of 0.11, i.e. the ratio of the mean IC_{50} for methionine-enkephalin in the mouse vas deferens to that in the guinea-pig ileum. The values for inhibition of [^{3}H] -naltrexone binding were multiplied by the conversion factor of 0.18, i.e. the ratio of the mean IC_{50} for methionine-enkephalin for inhibition of binding of [^{3}H] -leucine-enkephalin to that of binding of [^{3}H] -naltrexone.

enkephalin increases the potencies in both guinea-pig ileum and mouse vas deferens by factors of 14 and 7 respectively, without altering significantly the affinities to the [^{3}H]-naltrexone and [^{3}H]-leucine-enkephalin binding sites (table I). This effect is most likely due to a decrease in the enzymatic degradation of the peptide in the pharmacological models since the binding assays were carried out at 0–4°C. Replacement of L-Leu by D-Leu increased activity in the mouse vas deferens without a major change in the affinity to the binding sites; on the other hand, the activity in the guinea-pig ileum was reduced somewhat. The pharmacological pattern was still of the type characteristic of leucine-enkephalin, perhaps even to an exaggerated extent: the peptide was much more potent in the mouse vas deferens than in the guinea-pig ileum and the affinity for the [^{3}H]-leucine-enkephalin binding site was much higher than that for the [^{3}H]-naloxone binding site. This peptide has been shown (1) to have antinociceptive activity after injection into the cerebral ventricles. When the C-terminal leucine was replaced by amides of proline, the most important change was an increase in the activity in the guinea-pig ileum; there was also an increase in the affinity for the [^{3}H]-naltrexone binding site with a simultaneous loss in affinity for the [^{3}H]-leucine-enkephalin binding site (table I). Both compounds have antinociceptive activity after intravenous and subcutaneous injection as shown by *Székely et al.* (16). Another analogue, which is a powerful antinociceptive agent and shows activity even after oral administration as demonstrated by *Roemer et al.* (14), is Tyr-D-Ala-Gly-NCH_3Phe-Met(O)-ol. This compound shows a further shift in relative activities in favour of the guinea-pig ileum and the [^{3}H]-naltrexone binding sites. Its pharmacological pattern is very different from that of methionine-enkephalin and has become more similar to that of morphine which, however, has a much lower overall activity than the enkephalin analogue.

Alterations at the two terminal amino acid residues have considerable effects on the pharmacological pattern of the enkephalins. When the C-terminal leucine of D-Ala^2-L-Leu^5-enkephalin is decarboxylated, the relative potency is increased in the guinea-pig ileum and markedly decreased in the mouse vas deferens while the affinity for the naltrexone binding site is improved and that for the enkephalin binding site diminished (table II). When a methyl group is now introduced in the amino group of tyrosine, the potencies are decreased in all four assay systems but not to the same extent, those for the mouse vas deferens and the leucine-enkephalin binding site being particularly affected. In other words, the free carboxylic group at the C-terminus and the primary amino group at the N-terminus would appear to be essential for the maintenance of an enkephalin-like pharmacological pattern. When the leucine residue is removed from D-Ala^2-L-Leu^5-enkephalin, the resulting tetrapeptide is less active than the parent compound; the loss in activity is again much more pronounced for the mouse vas deferens and the leucine-enkephalin binding site, showing the importance of Leu^5 or Met^5 for the enkephalin-like properties of the pentapeptides (table II).

Table II. The effects of changes at the terminal amino acid residues on the relative potencies of D-Ala^2-leucine-enkephalin, referred to the potencies of methionine-enkephalin to inhibit the contractions of the mouse vas deferens and the binding of [^3H]-leucine-enkephalin in brain homogenates

Compound	Inhibition of contractions of			Inhibition of binding in brain homogenates of		
	Guinea-pig ileum	Mouse vas deferens	Gpi/Mvd	[^3H]-Nal-trexone	[^3H]-Leucine-enkephalin	Naltr/Leu-enk
Tyr-D-Ala-Gly-Phe-Leu	0.61	9.9	0.06	0.058	0.56	0.10
Tyr-D-Ala-Gly-Phe-NH$(CH_2)_2CH(CH_3)_2$	0.90	0.70	1.3	0.14	0.25	0.56
NCH_3 Tyr-D-Ala-Gly-Phe-NH$(CH_2)_2CH(CH_3)_2$	0.42	0.16	2.6	0.040	0.043	0.93
Tyr-D-Ala-Gly-Phe	0.074	0.064	1.2	0.020	0.024	0.83
Tyr-D-Ala-Gly-NH$(CH_2)_2$Ph	0.11	0.050	2.2	0.056	0.011	5.1
NCH_3 Tyr-D-Ala-Gly-NH$(CH_2)_2$Ph	0.064	0.026	2.5	0.027	0.002	13.5
Morphine	0.19	0.027	7.0	0.066	0.008	8.3
Methionine-enkephalin	0.11	1	0.11	0.18	1	0.18

The values are the means obtained as indicated in table I.

Antagonist Action of Naloxone against Different Enkephalin Analogues

It has been stressed (11) that the low effectiveness of naloxone against the action of the naturally occurring opioid peptides in the mouse vas deferens is strong supporting evidence for the view that the δ-receptors of the mouse vas deferens are different from the μ-receptors, with which the classical opiates interact. However, the opioid peptides can also interact with μ-receptors, which appear to be preponderant in the guinea-pig ileum where naloxone is equally effective against the opioid peptides and morphine. If this concept is correct, then naloxone should be a weaker antagonist in the mouse vas deferens against enkephalin analogues which retain their enkephalin-like pharmacological pattern, as for instance Tyr-D-Ala-Gly-Phe-D-Leu (K_e = 32 nM), than against enkephalin analogues which are more morphine-like, as for instance NCH_3Tyr-D-Ala-Gly-$NH(CH_2)_2$Ph (K_e = 6.3 nM) or Tyr-D-Ala-Gly-NCH_3Phe-Met(O)-ol (K_e = 5.7 nM). The values obtained so far are compatible with this concept.

Possible Relationships between Receptors and Function

Little is known about the functions that are mediated by the various receptors. However, it has been shown that, after injection into the cerebral ventricles of rats, D-Ala^2-D-Leu^5-enkephalin (Wellcome) which has a high affinity for δ-receptors has only 1% of the antinociceptive activity of Tyr-D-Ala^2-Gly-MePhe-Met(O)-ol (Sandoz) (2), whose affinity for the μ-receptors is as high as that of the Wellcome analogue to the δ-receptor. As a corollary, the affinity of the Sandoz compound to the δ-receptors is as low as that of the Wellcome compound to the μ-receptors (table I). It is therefore possible that the μ-receptors are more important for antinociceptive effects than the δ-receptors. β-Endorphin may owe its high antinociceptive potency to the fact that it binds equally well to μ-receptors and δ-receptors. This interpretation would, at least to some extent, explain the difficulty experienced by many observers in showing that naloxone has a hyperalgesic effect in normal animals.

Summary

The three agonists, methionine-enkephalin, leucine-enkephalin and β-endorphin have different pharmacological patterns. It may be of particular importance that they vary in their relative affinities to the enkephalin and naltrexone binding sites in the brain; the former are probably related to δ-receptors prevalent in the mouse vas deferens and the latter to μ-receptors prevalent in the guinea-pig. It is possible that μ-receptors are more important for the mediation of analgesic effects than δ-receptors. An understanding of the pharmacokinetics of the opioid peptides will be of basic importance for the design of enkephalin analogues suitable for use as analgesics in man.

References

1 Baxter, M.G.; Goff, D.; Miller, A.A., and Saunders, I.A.: Effect of a potent synthetic opioid pentapeptide in some antinociceptive and behavioural tests in mice and rats. Br. J. Pharmacol. *59:* 455–456 (1977).
2 Blasig, J. and Herz, A.: Personal communication (1978).
3 Bloom, F.; Battenberg, E.; Rossier, J.; Ling, N., and Guillemin, R.: Neurons containing β-endorphin in rat brain exist separately from those containing enkephalin: immunocytochemical studies. Proc. natn. Acad. Sci. USA *75:* 1591–1595 (1978).
4 Elde, R.; Hökfelt, T.; Johansson, O., and Terenius, L.: Immunohistochemical studies using antibodies to leucine-enkephalin: initial observations on the nervous system of the rat. Neuroscience *1:* 349–351 (1976).
5 Gillan, M.G.C.; Kosterlitz, H.W., and Paterson, S.J.: Comparison of the binding characteristics of tritiated opiates and opioid peptides. Br. J. Pharmacol. *66:* 86–87 (1979).
6 Hong, J.S.; Hang, H.-Y.T.; Fratta, W., and Costa, E.: Determination of methionine in discrete regions of rat brain. Brain Res. *134:* 383–386 (1977).
7 Hughes, J.; Kosterlitz, H.W., and Smith, T.W.: The distribution of methionine-enkephalin and leucine-enkephalin in the brain and peripheral tissues. Br. J. Pharmacol. *61:* 639–647 (1977).
8 Hutchinson, M.; Kosterlitz, H.W.; Leslie, F.M.; Waterfield, A.A., and Terenius, L.: Assessment in the guinea-pig ileum and mouse vas deferens of benzomorphans which have strong antinociceptive activity but do not substitute for morphine in the dependent monkey. Br. J. Pharmacol. *55:* 541–546 (1975).
9 Kosterlitz, H.W. and Hughes, J.: Development of the concepts of opiate receptors and their ligands; in Costa and Trabucchi, The endorphins. Advances in biochemical psychopharmacology; vol. 18, pp. 31–44 (Raven Press, New York 1978).
10 Kosterlitz, H.W.; McKnight, A.T.; Waterfield, A.A.; Gillan, M.G.C., and Paterson, S.J.: Assessment of analogues of opioid peptides in four parallel assays. Rep. 40th Meeting, Committee Probl. Drug Dependence, pp. 139–150 (1978).
11 Lord, J.A.H.; Waterfield, A.A.; Hughes, J., and Kosterlitz, H.W.: Endogenous opioid peptides: multiple agonists and receptors. Nature, Lond. *267:* 495–499 (1977).
12 Martin, W.R.: Opioid antagonists. Pharmac. Rev. *19:* 463–521 (1967).
13 Martin, W.R.; Eades, C.G.; Thompson, J.A.; Huppler, R.E., and Gilbert, P.E.: The effects of morphine- and nalorphine-like drugs in the nondependent and morphine-dependent chronic spinal dogs. J. Pharmac. exp. Ther. *197:* 517–532 (1976).
14 Roemer, D.; Buescher, H.H.; Hill, R.C.; Pless, J.; Bauer, W.; Cardinaux, F.; Closse, A.; Hauser, D., and Huguenin, R.: A synthetic enkephalin with prolonged parenteral and oral analgesic activity. Nature, Lond. *268:* 547–549 (1977).
15 Simantov, R.; Kuhar, J.J.; Uhl, G.R., and Snyder, S.H.: Opioid peptide enkephalins: immunohistochemical mapping in the rat central nervous system. Proc. natn. Acad. Sci. USA *74:* 2167–2171 (1977).
16 Székely, J.I.; Rónai, A.Z.; Dunai-Kovács, Z.; Miglécz, E.; Bertzétri, I.; Bajusz, S., and Gráf, L.: (D-Met^2, Pro^5)-Enkephalin amide: a potent morphine-like analgesic. Eur. J. Pharmacol. *43:* 293–294 (1977).
17 Watson, S.J.; Akil, H.; Richard III, C.W., and Barchas, J.D.: Evidence for two separate opiate peptide neuronal systems. Nature, Lond. *275:* 226–228 (1978).

Dr. H.W. Kosterlitz, Unit for Research on Addictive Drugs, University of Aberdeen, Aberdeen (UK)

Prog. biochem. Pharmacol., vol. 16, pp. 11–21 (Karger, Basel 1980)

Functional Aspects of Endorphins[1]

A. Herz, V. Höllt, R. Przewłocki, H. Osborne, Ch. Gramsch and T. Duka
Max-Planck-Institut für Psychiatrie, Department of Neuropharmacology, Munich

Introduction

The discovery of endogenous ligands for opiate receptors (endorphins) in brain and pituitary initiated efforts to evaluate the physiological significance of these peptides (30). Although there is little doubt that the endorphins present in brain and other nervous tissue may act as neurotransmitters or neuromodulators, there is little information as yet about their particular functional role in defined neuronal systems. Still less is known about the putative hormonal function of the endorphins present in the pituitary and the relationship between brain and pituitary endorphins. Some of these questions are evaluated in the present paper on the basis of studies of the distribution of endorphins in brain and pituitary and investigations of release of endorphins in animals and in man. In addition, a possible role of endorphins in opiate addiction is discussed in relation to the changes taking place in the endorphinergic system upon chronic opiate treatment.

Radioimmunoassays for Enkephalins and β-Endorphin

Methionine-enkephalin (met-enkephalin), leucine-enkephalin (leu-enkephalin) and β-endorphin were determined by radioimmunoassays. The antisera were generated in rabbits using thyroglobulin-coupled peptides as immunogens. The antisera directed against met-enkephalin and leu-enkephalin showed high specificity: the cross-reactivity was about 0.3% for leu-enkephalin with antiserum against met-enkephalin and for met-enkephalin with antisera against leu-enkephalin. β-Endorphin, β-lipotropin (β-LPH), β-LPH_{61-69}, β-LPH_{61-76}, and β-LPH_{61-77} did not cross-react with the enkephalin antisera. The detection limit

[1] Supported by a grant from the Deutsche Forschungsgemeinschaft, Bonn–Bad Godesberg.

for met-enkephalin and leu-enkephalin was below 40 fmoles/tube. The antiserum directed against human β-endorphin displayed high avidity (detection limit below 3 fmoles/tube) and did not cross-react with met-enkephalin, leu-enkephalin, α-endorphin or γ-endorphin. There was, however, a 50% cross-reactivity (on a molar basis) with human β-lipotropin. For details of these assays see *Wesche et al.* (32) and *Höllt et al.* (14).

β-Immunoreactivity was further characterized by gel-filtration using a Sephadex G-50 superfine column equilibrated and eluated with a buffer system previously described by *Guillemin et al.* (10). For details see *Przewłocki et al.* (26).

Distribution of Endorphins in Brain

The highest concentrations of leu- and met-enkephalin in rat brain are found in the corpus striatum and in the hypothalamus, whereas rather low concentrations are present in the hippocampus and the cerebellum (32, 33). The highest concentrations of β-endorphin, on the other hand, are found in the hypothalamus, and it is not present in the striatum (2). Similar results have been obtained in our laboratory for human brain with regard to the distribution of β-endorphin and enkephalin. Besides the hypothalamus, significant concentrations of β-endorphin were detected in the corpus mamillare and in various parts of the thalamus and the midbrain, but not in the striatum of humans (8, 9) (fig. 1).

Column chromatographic studies have recently revealed that both in rat and in human brain most of the β-endorphin immunoreactivity is due to β-endorphin; β-LPH is present, if at all, only at low concentrations (*Höllt and Gramsch,* unpublished). These findings are consistent with immunocytochemical studies which have shown β-endorphin-containing cell bodies present in the basal hypothalamus which give rise to the fibres projecting through the anterior hypothalamus area to the medial dorsal thalamus and midbrain periaqueductal gray (2). In contrast, enkephalin-containing cell bodies and fibres are present in many regions of the CNS, including the spinal cord, with the highest density in the globus pallidus (6).

Distribution of Endorphins in the Pituitary Gland

The highest amounts of β-endorphin immunoreactive material in rats and humans are found in the pituitary. In rats, a species in which the intermediate lobe is clearly separable anatomically, this part contains the highest concentration. High levels are also found in the anterior lobe where it is localized in distinct cells, while the posterior lobe contains much less (5). There is some

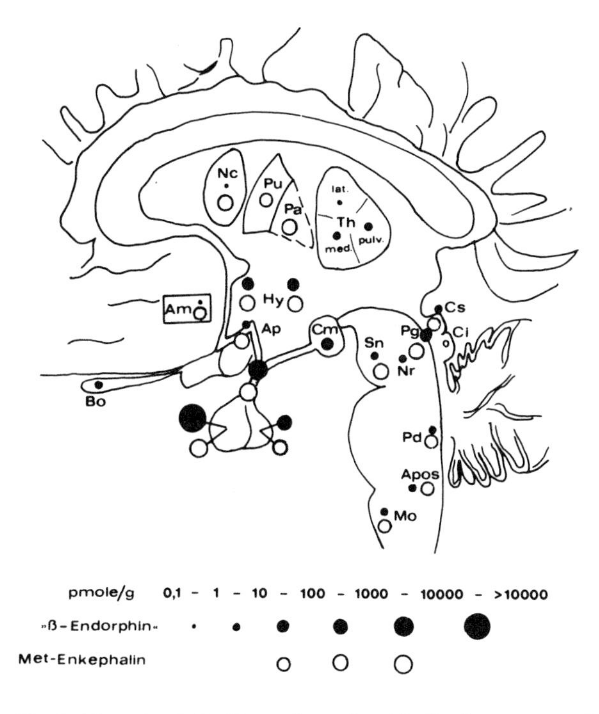

Fig. 1. Met-enkephalin-like and β-endorphin-like immunoreactivity measured in various parts of human brain: Am = Amygdala; Ap = Area preoptica hypothalami; Apos = Area postrema; Bo = Bulbus olfactorius; Cm = Corpus mamillare; Cs = Colliculi superiores; Ci = Colliculi inferiores; Hy = Hypothalamus; Mo = Medulla oblongata dorsalis; Nc = Nucleus caudatus; Nr = Nucleus ruber; Pa = Pallidum; Pd = Pons dorsalis; Ps = Periaqueductal gray; Pn = Putamen; Th = Thalamus, lateral thalamus, medial thalamus, pulvinar thalamus; Sn = Substantia nigra.

discrepancy concerning the nature of this β-endorphin immunoreactivity in the pituitary. It has been pointed out (19) that immunoreactivity in the anterior lobe is exclusively due to β-LPH, and that β-endorphin might be artificially generated from β-LPH by the extraction procedures employed. However, in our experiments, in which anterior lobes of rat pituitaries were extracted after boiling the intact tissue with 0.1 N HCl in order to minimize enzymatic breakdown, chromatographic separation of immunoreactive components showed that the amounts of β-endorphin and β-LPH were about equal. In the intermediate/posterior lobe, however, almost all immunoreactive material proved to be β-endorphin (25).

Also under discussion is the question of whether enkephalins occur in the pituitary gland. Whereas various investigators have found no enkephalins or only very small amounts, we recently detected relatively high amounts of met- and leu-enkephalin using highly specific antibodies for both peptides (5). The distribution pattern of the enkephalins was quite different from that of β-endorphin, and significant amounts were found in the pars nervosa. These findings lead to the speculation that enkephalins are the natural ligands of the opiate receptors found in the posterior lobe of the pituitary (27).

Release of Endorphins from Brain Tissue in Vitro

Endorphins can be released from brain tissue *in vitro* in response to depolarizing agents. Several investigators have reported the release of enkephalins from striatal and globus pallidus slices (11, 15, 22). Both enkephalins are obviously released in the same proportion as they are present in the tissue (1, 22). Recently we succeeded in demonstrating release of β-endorphin immunoreactive materials from rat hypothalamic slices by potassium (23). Both enkephalin- and β-endorphin-release proved to be calcium-dependent, a finding which is similar to that reported for the release of putative neurotransmitters in the CNS and as such may be taken as support for the view that endorphins have neurotransmitter or neuromodulator functions.

Release of Endorphins from Pituitary in Vitro

The release of immunoreactive β-endorphin from isolated pituitary lobes has been studied by our group (25, 26) and from isolated pituitary cells by *Vale et al.* (31). The experiments revealed a clear difference between the anterior and the intermediate/posterior lobe. Increase of potassium ions induced release of β-endorphin immunoreactive material from the anterior lobe, but not from the intermediate/posterior lobe. Lysine-vasopressin, hypothalamic extract and noradrenaline also released β-endorphin immunoreactive material from the anterior lobe, an effect which could be antagonized by corticosterone. On the other hand, dopamine and the ergot alkaloid ergonovine blocked the spontaneous release of β-endorphin from the intermediate/posterior lobe, but not from the anterior lobe. The chromatographic separation of the immunoreactive components revealed that 30–40% of the material released from the anterior lobe was due to β-endorphin, while the material spontaneously released from the intermediate/posterior lobe was almost exclusively β-endorphin. Similar results have been reported by *Vale et al.* (31) using cell cultures obtained from anterior lobe cells of rat pituitaries. There was a general parallel between the release of β-endorphin and ACTH, a result which may be explained by the finding that β-endorphin and ACTH are contained in the same cells and originate from the same precursor molecule (21). Interestingly, in our experiments with intact an-

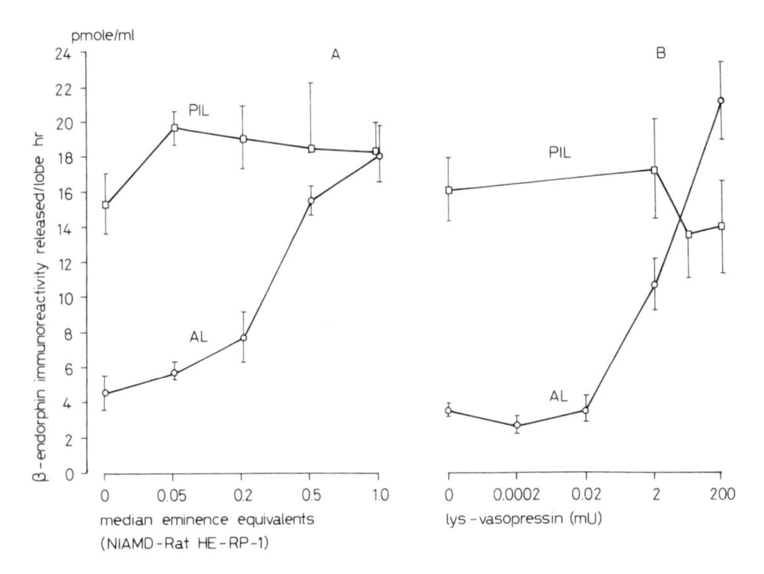

Fig. 2. Release of β-endorphin immunoreactive material from anterior lobe (AL) or intermediate/posterior lobe (PIL) of rat pituitary by median eminence extract or lysine-vasopressin *in vitro.*

terior lobes vasopressin proved to be more effective in promoting release than in the cultured cells. Taking these results together, the mechanisms of release from the different lobes of the pituitary seem to be quite different. β-Endorphin and β-LPH are released from the anterior pituitary lobe in response to releasing factors (probably closely related to vasopressin) into the blood stream; the mode of release of β-endorphin from the intermediate/posterior lobe cells is much more obscure (fig. 2).

Release of β-Endorphin in Vivo

From the *in vitro* data it may be expected that β-endorphin is released concomitantly with ACTH from the pituitary during stress. This has indeed been shown by *Guillemin et al.* (10), who also described increased β-endorphin levels in rat plasma after adrenalectomy and decreased levels after dexamethasone treatment. These results are supplemented by findings which showed increased β-endorphin immunoreactivity in plasma after insulin and vasopressin as well as after blockage of cortisol synthesis by metyrapone in rats (12). These data are a further indication that cortisol regulates the release of β-endorphin via feed-back mechanisms, and may indicate that the physiological regulation of β-endorphin and ACTH is under the influence of the same factors.

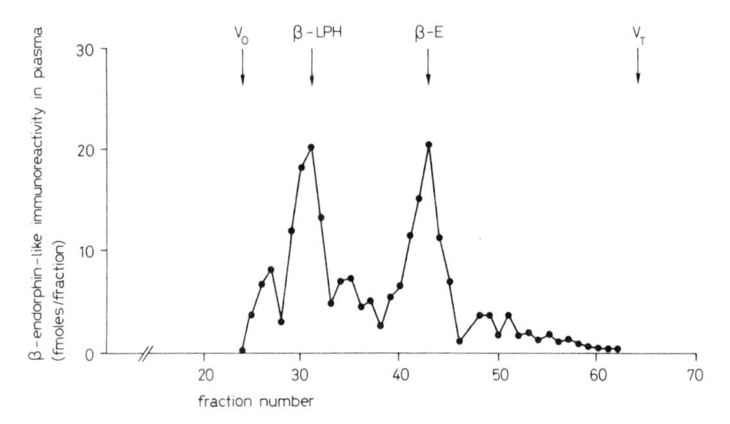

Fig. 3. β-Endorphin-like immunoreactivity in human plasma. 30 ml plasma was extracted with silicic acid according to *Höllt et al.* (14) and subjected to gel-filtration on a Sephadex G-50 column. Vo = void volume; Vt = total volume.

The concomitant release of ACTH and β-endorphin from the pituitary of humans has recently been questioned by *Suda et al.* (29), who could not detect significant amounts of β-endorphin in human plasma. Only under pathological conditions, such as Nelson's disease, did these authors find measurable amounts of β-endorphin and β-LPH. In view of the fact that β-endorphin, but not β-LPH, represents a ligand of the opiate receptors, the distinction between these peptides seems to be essential. We were recently able to show by gel chromatography that β-endorphin is present in normal human plasma (*Höllt,* unpublished). β-Endorphin and β-LPH were found in more or less similar though rather low amounts (fig. 3).

Increased levels of β-endorphin immunoreactive material were recently found in the plasma of patients with Addison's disease, Nelson's disease and Cushing's disease. There was an excellent correlation between levels of β-endorphin immunoreactive material and ACTH. Varying proportions of β-LPH and β-endorphin were detected in these patients using gel chromatography, β-LPH usually predominating (12; *Höllt,* unpublished).

The functional significance of β-endorphin released from the pituitary is still unknown. It may be speculated that peripheral organs, e.g. the gastrointestinal tract, represent a target for β-endorphin. The strong gastrointestinal effects of opiates and the occurrence of opiate receptors and of enkephalin in the gut support such a view. But there is so far little direct evidence. Results obtained recently by our group showing that the intestines of fetal guinea pigs respond to naloxone by a withdrawal-like contracture (while preparations obtained from adult guinea pigs do not) suggest an involvement of endorphins in pregnancy

(17), although this does not necessarily point to β-endorphin. These results may be related to the recent finding of increased β-endorphin levels in maternal blood during labour. Though this may be a simple 'stress phenomenon', it is interesting that β-endorphin has been also dectected in the fetal blood after delivery (18; *Csontos and Höllt,* unpublished).

Another possibility to be considered as a target organ for β-endorphin released from the pituitary into blood is the brain. It is doubtful, however, whether significant amounts of β-endorphin can enter the brain via blood flow. The data available on the central effects of systemically applied β-endorphin are somewhat contradictory (10, 20). Blood levels observed under stress are orders of magnitude lower than those necessary to induce visible central effects after systemic injection. Data showing a rapid permeation of β-endorphin into the CSF of rabbits (24) could not be reproduced in our laboratory. Nevertheless, the question of whether the brain is a target of β-endorphin released from the pituitary into circulation needs further investigation.

Endorphins and Opiate Addiction

The mechanisms underlying opiate addiction are still unknown. It was therefore not surprising that soon after the detection of the endorphins their possible role in opiate addiction was examined by several investigators. By analogy to other neurotransmitter systems *Kosterlitz and Hughes* (16) postulated that prolonged opiate treatment might cause feed-back inhibition of endorphin synthesis; this may result in a deficiency of endorphins, accounting perhaps for the appearance of withdrawal signs. However, various investigations did not reveal significant changes in the enkephalin content of various areas of rat brain after (conventional) chronic morphine treatment or after naloxone-induced precipitated withdrawal (3, 7, 32). In view of the fact that levels give little information about the functional state of a neurotransmitter system, these negative results do not exclude the possibility that tolerance/dependence is associated with changes in enkephalin metabolism. In this connection, however, it is of interest that no significant changes could be detected in the release of enkephalin from striatal slices following naloxone-induced precipitated withdrawal (*Osborne,* unpublished).

Recent experiments revealed changes in β-endorphin after acute and chronic morphine treatment in rats. High doses of morphine (50 mg/kg) induced a significant increase in β-endorphin immunoreactive material in the plasma while the β-endorphin content of the anterior and intermediate/posterior pituitary lobes remained unchanged. After pellet implantation for 10 days plasma and pituitary β-endorphin immunoreactivity were unaltered. Naloxone-precipitated withdrawal, however, greatly increased β-endorphin immunoreactivity in the plasma. At the same time β-endorphin immunoreactivity in the anterior pituitary lobe was reduced, but remained constant in the intermediate/posterior lobe. This release is obviously mediated by the hypothalamus since no change was observed in *in vitro* experiments on the release of β-endorphin from pituitaries of tolerant/dependent rats following naloxone treatment. At present it is not clear whether this release of β-endorphin immunoreactivity into blood is due to a specific receptor-coupled mechanism or to the non-specific stress effects of withdrawal. In view of the finding that corticosteroids in plasma increase during withdrawal, and ACTH and β-endorphin are concomitantly released from the pituitary during stress, the latter explanation seems more probable (13).

Effects which seem to be more specific were observed when the time of exposure of rats to morphine was prolonged to one month, and the brain as well as the pituitary was analysed for enkephalins and β-endorphin. After such treatment a significant decrease of the enkephalin content in the striatum and of β-endorphin in the septum and the midbrain were observed, while the β-endorphin content in the hypothalamus remained unchanged. In the pituitary the enkephalin content as well as the β-endorphin content in the intermediate/posterior lobe was found to be decreased by 60%, but the β-endorphin content in the anterior lobe was unchanged. One explanation for these changes might be that long-term treatment inhibits endorphin synthesis, inducing a deficiency in these peptides as originally postulated by *Kosterlitz and Hughes* (16). The fact that endorphin levels are decreased only in some brain areas and in the intermediate/posterior lobe, but not in other brain areas or in the anterior pituitary lobe, speaks against the possibility that we are dealing with non-specific 'toxic' effects of morphine.

Changes in endorphins could play a role in the subtle protracted signs of abstinence known to persist for a long time after cessation of the more dramatic acute withdrawal signs. Even before the endorphins were discovered, *Dole and Nyswander* (4) suggested that some type of metabolic disease may underlie narcotic addiction. It now seems reasonable to suggest that such changes could take place in the endorphinergic system. The few data available so far on endorphin metabolism (28) (pointing to a very low turnover rate) may explain why it needs a long time of opiate exposure to visualise such an effect by measuring endorphin levels.

Summary

Radioimmunoassay of methionine-enkephalin, leucine-enkephalin and β-endorphin were used in order to study the distribution and release of endorphins. The distribution pattern of enkephalin immunoreactivity in brain, including human brain, is quite different from that of β-endorphin immunoreactivity. Separation of β-endorphin and β-lipotropin by column chromatography revealed that the contribution of β-lipotropin to β-endorphin immunoreactivity in brain is very small. In the anterior lobe of the pituitary both β-endorphin and β-lipotropin were found, whereas in the intermediate/posterior lobe almost all immunoreactivity was due to β-endorphin; considerable amounts of enkephalin were also detected. Raising the concentration of potassium ions stimulated the release of met- and leu-enkephalin from striatal slices and the release of β-endorphin immunoreactive material(s) from hypothalamic slices; both phenomena were dependent upon the presence of calcium ions. Studies of the release of β-endorphin from isolated rat pituitaries revealed characteristic differences between the anterior and intermediate/posterior lobes; e.g., lysine vasopressin and extracts from the median eminence were highly effective in releasing β-endorphin from the anterior lobe without affecting the release from the intermediate/posterior lobe; on the other hand, dopamine inhibited β-endorphin release from the intermediate/posterior lobe without affecting release from the anterior lobe. Increased β-endorphin levels were found after various stress conditions in rat plasma, as well as after treatment with metyrapone and vasopressin. In normal human plasma significant amounts of β-endorphin were detected; increased levels were found in Addison's, Nelson's and Cushing's disease.

Chronic opiate treatment of rats for 10 days did not affect brain levels of enkephalin or the β-endorphin content of the hypothalamus, pituitary and plasma. Precipitated withdrawal decreased β-endorphin in the anterior lobe and hypothalamus and increased β-endorphin levels in the plasma. Long-term morphine treatment (30 days) decreased enkephalin and β-endorphin content in some brain areas and in the intermediate/posterior pituitary lobe but not in the anterior lobe.

References

1 Bayon, A.; Rossier, J.; Mauss, A.; Bloom, F.; Iversen, L.; Ling, N., and Guillemin, R.: *In vitro* release of (5-methionine)-enkephalin and (5-leucine)-enkephalin from rat globus pallidus. Proc. natn. Acad. Sci. USA *75:* 3503–3506 (1978).

2 Bloom, F.; Battenberg, E.; Rossier, J.; Ling, N., and Guillemin, R.: Neurons containing β-endorphin in rat brain exist separately from those containing enkephalin: Immunocytological studies. Proc. natn. Acad. Sci. USA *75:* 1991–1995 (1978).

3 Childers, S.R.; Simantov, R., and Snyder, S.H.: Enkephalin: Radioimmunoassay and radioreceptor assay in morphine dependent rats. Eur. J. Pharmacol. *46:* 289 (1977).

4 Dole, V.P. and Nyswander, M.: Heroin-addiction – a metabolic disease. Archs int. Méd. exp. *120:* 19–24 (1967).

5 Duka, Th.; Höllt, V.; Przewłocki, R., and Wesche, D.: Distribution of methionine- and leucine-enkephalin within the rat pituitary gland measured by highly specific radioimmunoassays. Biochem. biophys. Res. Commun. *85:* 1119–1127 (1978).

6 Elde, R.; Hökfelt, T.; Johannson, O., and Terenius, L.: Immunohistochemical studies using antibodies to leucine-enkephalin: Initial observations on the central nervous system of the rat. Neuroscience *1:* 349–355 (1976).

7 Fratta, W.; Yang, H.Y.; Hong, J., and Costa, E.: Stability of met-enkephalin content in brain structures of morphine-dependent or foot shock stressed rats. Nature, Lond. *268:* 452–453 (1977).

8 Gramsch, Ch.; Höllt, V.; Mehraein, P.; Pasi, A., and Herz, A.: Regional distribution of endorphins in human brain and pituitary; in van Ree and Terenius, Characteristics and function of opioids, pp. 277–278 (Elsevier/North-Holland Biomedical Press, Amsterdam 1978).

9 Gramsch, Ch.; Höllt, V.; Mehraein, P.; Pasi, A., and Herz, A.: Regional distribution of methionine-enkephalin and β-endorphin-like immunoreactivity in human brain and pituitary. Brain Res. *171:* 261–270 (1979).

10 Guillemin, R.; Vargo, T.; Rossier, J.; Minick, S.; Ling, S.; Rivier, C.; Vale, M., and Bloom, F.: β-Endorphin and adrenocorticotropin are secreted concomitantly by pituitary gland. Science, N.Y. *197:* 1367–1369 (1977).

11 Henderson, G.; Hughes, J., and Kosterlitz, H.W.: *In vitro* release of leu- and met-enkephalin from the corpus striatum. Nature, Lond. *271:* 677–679 (1978).

12 Höllt, V.; Emrich, H.M.; Müller, O.A., and Fahlbusch, R.: β-Endorphin-like immunoreactivity (β-ELI) in human plasma and cerebrospinal fluid; in van Ree and Terenius, Characteristics and function of opioids, pp. 279–280 (Elsevier/North-Holland Biomedical Press, Amsterdam 1978).

13 Höllt, V.; Przewłocki, R., and Herz. A.: β-Endorphin-like immunoreactivity in plasma, pituitaries and hypothalamus of rats following treatment with opiates. Life Sci. *23:* 1057–1066 (1978).

14 Höllt, V.; Przewłocki, R., and Herz. A.: Radioimmunoassay of β-endorphin. Basal and stimulated levels in extracted rat plasma. Naunyn-Schmiedeberg's Arch. Pharmacol. *303:* 171–174 (1978).

15 Iversen, L.L.; Iversen, S.O.; Bloom, F.E.; Vargo, T., and Guillemin, R.: Release of enkephalin from rat globus pallidus *in vitro.* Nature, Lond. *271:* 679–681 (1978).

16 Kosterlitz, H.W. and Hughes, J.: Some thoughts on the significance of enkephalin, the endogenous ligands. Life Sci. *17:* 91–96 (1975).

17 Kromer, W. and Teschemacher, H.: An opiate withdrawal-like phenomenon in the fetal guinea pig ileum upon naloxone challenge. Eur. J. Pharmacol. *49:* 445–446 (1978).

18 Kromer, W.; Teschemacher, H.; Fischer, C.; Höllt, V.; Schulz, R., and Voigt, K.H.: Indication for a possible role of endorphins in pregnancy; in van Ree and Terenius, Characteristics and function of opioids, pp. 281–282 (Elsevier/North-Holland Biomedical Press, Amsterdam 1978).

19 Liotta, A.S.; Suda, T., and Krieger, D.T.: β-Lipotropin is the major opioid-like peptide of human pituitary and rat pars distalis: Lack of significant β-endorphin. Proc. natn. Acad. Sci. USA *75:* 2950–2954 (1978).

20 Loh, H.H.; Tseng, L.F.; Wei, E., and Li, C.H.: β-Endorphin is a potent analgetic agent. Proc. natn. Acad. Sci. USA *73:* 2895–2898 (1976).

21 Mains, R.E.; Eipper, B.A., and Ling, N.: Common precursor to corticotropins and endorphins. Proc. natn. Acad. Sci. USA *74:* 3014–3018 (1977).

22 Osborne, H.; Höllt, V., and Herz, A.: Potassium-induced release of enkephalins from rat striatal slices. Eur. J. Pharmacol. *48:* 219–227 (1978).

23 Osborne, H.; Przewłocki, R.; Höllt, V., and Herz, A.: Release of β-endorphin-like immunoreactivity from rat hypothalamus *in vitro.* Eur. J. Pharmacol. *55:* 425–428 (1979).

24 Pezalla, P.D.; Lis, M.; Seidah, N.G., and Chrétien, M.: Lipotropin, melanotropin and endorphin *in vivo* catabolism and entry into cerebrospinal fluid. J. Can. Sci. Neurol. *3:* 183–188 (1978).

25 Przewłocki, R.; Höllt, V., and Herz, A.: Release of β-endorphin from rat pituitary *in vitro.* Eur. J. Pharmacol. *51:* 179–181 (1978).

26 Przewłocki, R.; Höllt, V.; Voigt, K.H., and Herz, A.: Distinctive *in vitro* release of β-endorphin from the anterior compared to the posterior/intermediate lobe of the pituitary. Life Sci. *24:* 1601–1608 (1979).

27 Simantov, R. and Snyder, S.H.: Brain pituitary opiate mechanisms: Pituitary opiate receptor binding, radioimmunoassays for methionine, enkephalin and leucine-enkephalin and ^{3}H-enkephalin interactions with the opiate receptor; in Kosterlitz, Opiates and endogenous opioid peptides, pp. 41–48 (Elsevier/North-Holland Biomedical Press, Amsterdam 1976).

28 Sosa, R.P.; Knight, A.T.; Hughes, J., and Kosterlitz, H.W.: Incorporation of labelled amino acids into enkephalins. FEBS Lett. *84:* 195–198 (1977).
29 Suda, T.; Liotta, A.S., and Krieger, D.T.: β-Endorphin is not detectable in plasma from normal human subjects. Science, N.Y. *202:* 221–223 (1978).
30 Teschemacher, H.: Endogenous ligands of opiate receptors (endorphins); in Herz, Developments in opiate research, pp. 67–151 (Marcel Dekker, New York 1978).
31 Vale, W.; Rivier, C.; Yang, L.; Minick, S., and Guillemin, R.: Effects of purified hypothalamic corticotropin-releasing factor and other substances on the secretion of adrenocorticotropin and β-endorphin immunoreactivities *in vitro.* Endocrinology *103:* 1911–1915 (1978).
32 Wesche, D.; Höllt, V., and Herz, A.: Radioimmunoassays of enkephalins. Regional distribution in rat brain after morphine treatment and hypophysectomy. Arch. Pharmakol. *301:* 79–82 (1977).
33 Yang, H.Y.; Hong, J.S., and Costa, E.: Regional distribution of leu- and met-enkephalin in rat brain. Neuropharmacology *16:* 303–307 (1977).

Dr. A. Herz, Max-Planck-Institut für Psychiatrie,
Department of Neuropharmacology, D–8000 Munich (FRG)

Prog. biochem. Pharmacol., vol. 16, pp. 22–31 (Karger, Basel 1980)

Morphine-Like Peptides: Their Regulation in the Neuroendocrine System and the Effect of Guanyl Nucleotides and Divalent Ions on Opiate Receptor Binding[1]

R. Simantov

Department of Genetics, The Weizmann Institute of Science, Rehovot

Introduction

It is conceivable that several major *in vivo* effects of opiates, e.g. euphoria and regulation of pain, are initiated or assembled in the central nervous system. On the other hand, studies over the years have shown that opiates significantly alter the normal metabolism of several hormones (2, 8, 20, 23), including hormones secreted from either the posterior or the anterior lobes of the pituitary gland. Thus, acute treatment with morphine stimulates release of anti-diuretic hormone (ADH) from the posterior lobe and of prolactin and growth hormone from the anterior lobe. In male rats, morphine suppresses release of luteinizing hormone (LH) and hence also of testosterone (6). The hormonal system of female rats also shows sensitivity to acute morphine treatment. Interestingly, injection of morphine during a critical period on the day of proestrus inhibits ovulation (2). In experimental animals, acute morphine treatment also stimulates release of adrenocorticotropic hormone (ACTH) and adrenal steroids (28). It is not yet known if the effect of morphine on some or all of these hormones is regulated through the hypothalamus or acts directly on the pituitary gland, as was recently suggested for some hormones (33). The discovery of opioid peptides in the pituitary gland (7, 14, 15, 33) raises the possibility that these peptides might regulate some of these hormones under normal physiological conditions. In fact, studies conducted in the last two years have shown that endorphins, enkephalins, or 'pure' opiate antagonists like naloxone can stimulate the release of prolactin (4, 9, 27) and growth hormone (4, 27) and alter the release of luteinizing hormone releasing hormone (LHRH) (25). It has been

[1] The author is indebted to Mr. *A. Goldenberg* for excellent technical assistance and to Prof. *L. Sachs* for support and interest. *R.S.* is an incumbent of the C.S. Koshland Career Developmental Chair.

indicated recently that both pituitary endorphins, which are localized in the intermediate and the anterior lobe (33), and ACTH are synthesized from a common prohormone (18) and that they coappear in the circulation after stress (10). Both these peptides are released from anterior pituitary cells after depolarization with high concentrations of potassium in a calcium-dependent fashion (12, 22, 29). This study was undertaken in order to reveal what factors regulate synthesis of endorphins by pituitary cells. An *in vitro* rather than *in vivo* approach was used to avoid non-direct effects. We have also tested the binding of enkephalin to opiate receptors in different brain regions and the effect of guanyl nucleotides and divalent ions on this interaction.

Materials and Methods

AtT/20 mouse pituitary cells were cultured as described (29) except that the fetal calf serum (GIBCO, New York) was dialyzed for 4 days against 50 volumes (each day) of 0.15 N NaCl. Cells were seeded for the experiments at 5×10^5 cells/5 ml Eagle's culture medium containing 4× vitamins and amino acids and 10% dialyzed serum. On the day of inoculation the cultures were treated with 5 μl of ethyl alcohol containing various hormones to give the indicated final hormone concentration. Control dishes received 5 μl of ethyl alcohol. The cells with the culture medium were collected 96 h after addition of the hormones, and the plates washed with 3 ml phosphate-buffered saline, pH 7.4, which was added to the medium containing the cells. This mixture was divided into two parts and the medium was separated from the cells by low-speed centrifugation (500 g). One part of each pair was taken for counting of cells, and the other part for extraction of endorphins. For this, the cells were homogenized in 0.1 N HCl, the homogenate neutralized with 0.5 N Tris base, centrifuged for 10 min at 40,000 g, and the supernatant as well as aliquots of the culture medium were stored at −20°C until assayed. Endorphin activity was quantitated by the opiate radioreceptor assay as described (29), using aliquots of 50–100 μl. Data were calculated from a standard curve of methionine enkephalin.

Binding of ^{3}H-D-ala^2-methionine enkephalin (D-ala^2-(tyrosyl-3,5-^{3}H) enkephalin 5-L-methioninamide, Radiochemical Centre, 30 Ci/mmole) to various rat brain regions was conducted as described (29). Guanyl nucleotides were purchased from the Sigma Company.

Results

Effect of Glucocorticoid Hormones on Cellular Content and on Release of Endorphins by Pituitary Cells

It has been recently shown that the AtT/20 tumor cells, derived from mouse anterior pituitary, indicate both basal and potassium stimulated release of endorphins (29). As part of an attempt to identify neurotransmitters or hormones that might participate in regulation of endorphin synthesis, the AtT/20 cells were cultured with different steroids. Table I shows that the synthetic glucocorticoids dexamethasone and prednisolone indeed had dramatic effects on the release of

Table I. Effect of steroids on the cellular content and on release of endorphins from pituitary cells

Steroid	Concentration (nM)	Endorphins activity[1]		Cell growth cells/plate ($\times 10^6$)
		cellular	released	
None	–	1,025 ± 145	90 ± 16	6.8 ± 0.6
Dexamethasone	1.0	535 ± 100	35 ± 10	6.6 ± 0.8
	10.0	110 ± 55	20 ± 8	6.5 ± 0.8
	100.0	75 ± 40	15 ± 10	6.9 ± 0.6
Prednisolone	1.0	720 ± 65	75 ± 20	6.6 ± 0.7
	10.0	220 ± 35	38 ± 15	6.5 ± 0.8
	100.0	166 ± 30	25 ± 14	7.0 ± 1.0
Progesterone	1.0	1,205 ± 350	95 ± 22	7.3 ± 1.0
	10.0	1,180 ± 260	95 ± 16	7.0 ± 1.2
	100.0	960 ± 160	82 ± 20	6.7 ± 0.8
Estradiol	10.0	1,105 ± 160	86 ± 15	6.7 ± 1.0
	100.0	1,160 ± 225	96 ± 10	7.0 ± 1.1
Testosterone	10.0	1,140 ± 117	86 ± 20	6.9 ± 1.0
	100.0	1,031 ± 125	98 ± 15	6.5 ± 1.0
17-α-Epitestosterone	100.0	1,275 ± 178	103 ± 14	6.8 ± 0.7
Δ^4-Androsten-11β-3,17-dione	10.0	1,040 ± 130	96 ± 19	6.9 ± 0.5
	100.0	1,074 ± 125	82 ± 18	6.4 ± 0.7

[1] Endorphins activity determined as described in *Methods*. Data are pmoles methionine enkephalin equivalents/10^6 cells.

endorphins into the culture medium. These hormones at nanomolar concentrations significantly decreased the amount of endorphins released. However, they also caused a significant decrease in the cellular endorphin content (table I). The inhibitory effect of dexamethasone on the cellular levels is dose dependent with half maximal effect of nM levels. The specificity of the effect of these glucocorticoids was indicated by the following results: a) Estradiol benzoate, testosterone, progesterone, 17-α-epitestosterone and Δ^4-androsten-11β-3,17-dione had no such effect (table I); b) ten times higher concentration of estradiol benzoate and progesterone, or 100 times of Δ^4-androsten-11β-3,17-dione, did not suppress the effect of 1 nM dexamethasone (table II).

The possibility that decrease in the cellular content of endorphins might be an outcome of inhibition of cell proliferation was also investigated. Table I shows that proliferation of the AtT/20 cells was not affected by either of the steroids tested at the indicated concentrations.

Table II. Specificity of the effect of steroids on cellular content of endorphins

Steroid	Concentration (nM)	Cellular endorphin activity	
		without dexamethasone	with 1 nM dexamethasone
None	–	1,085 ± 160	370 ± 68
Estradiol	5.0	1,150 ± 210	360 ± 80
	10.0	1,100 ± 140	340 ± 65
Progesterone	5.0	1,096 ± 136	328 ± 60
	10.0	1,175 ± 170	340 ± 70
Δ^4-Androsten-11β-3,17-dione	100.0	1,030 ± 144	345 ± 50

Endorphins activity determined as described in table I.

We were also interested to discover whether the effects of dexamethasone and prednisolone were mediated by stimulation of specific receptors. Cells were incubated at 37°C with ^{3}H-dexamethasone in the presence and absence of different concentrations of non-labeled steroids (table III). Dexamethasone and prednisolone showed dose-dependent activity starting at 10 nM, and in concentrations as low as 30 nM they inhibited 56% and 40% respectively of ^{3}H-dexamethasone incorporation into AtT/20 nuclei. Steroids like progesterone and estradiol showed inhibition of only 5–20% at 100 nM, whereas cortisone, 17-α-epitestosterone and Δ^4-androsten-11β-3,17-dione were inactive at 10–100 nM and showed little effect at 1,000 nM (table III).

Effect of Guanyl Nucleotides and Divalent Ions on Binding of ^{3}H-D-Ala^2-Methionine Enkephalin Binding to Opiate Receptor

Studies conducted by *Blume* (3) have recently shown that GTP and its rather stable analogue GMP-PNP inhibit binding of opiates to opiate receptors. Later it was shown that these nucleotides have differential effects on opiate agonists and antagonists (5). In the present work we studied the effect of GMP-PNP on binding of the potent enkephalin analogue ^{3}H-D-ala^2-methionine enkephalin to opiate receptors of different regions of rat brain. GMP-PNP inhibits the binding of this peptide in a dose-dependent manner in all the regions tested (unpublished results). However, different brain regions show different sensitivity to this nucleotide (table IV). Thus, opiate receptors of the cerebral cortex, caudate nucleus and hypothalamus are much less affected than receptors of other brain regions such as the midbrain and medulla oblongata.

Figure 1 shows that inhibition of ^{3}H-D-ala^2-methionine enkephalin binding to brain opiate receptors by GMP-PNP is highly sensitive to the presence of some

Table III. Incorporation of ^{3}H-dexamethasone to nuclei of pituitary cells: inhibition by various steroids

Non-labeled steroid	Concentration (nM)	% Inhibition of ^{3}H-dexamethasone incorporation[1]
–	–	0
Dexamethasone	10	22 ± 5
	30	56 ± 12
	100	75 ± 10
Prednisolone	10	10 ± 6
	30	40 ± 10
	100	50 ± 15
Progesterone	10 or 30	0 ± 10
	100	5 ± 10
Estradiol	10 or 30	0 ± 15
	100	20 ± 5
Cortisone	10, 30 or 100	0 ± 15
	1,000	23 ± 10
17-α-Epitestosterone	100 or 1,000	0 ± 5
Δ^4-Androsten-11β-3,17-dione	100	0 ± 10
	1,000	12 ± 5

[1] One hundred percent incorporation was equal to 42 fmole ^{3}H-dexamethasone incorporated into 10^7 cells. The experiment was conducted as follows. Cells were dispersed in 0.1% EDTA in phosphate-buffered saline, pH 7.4, counted, washed with culture medium without serum and dispersed at 2×10^7 cells/ml of fresh medium. Then, cells were incubated for 40 min at 37°C in a humidified incubator with 4×10^{-8} M ^{3}H-dexamethasone (23 Ci/mmole, Radiochemical Centre) with or without the indicated concentrations of non-labeled steroids. Nuclei were prepared by freezing and thawing. Ten μM dexamethasone was taken as the baseline of maximal displacement.

Table IV. Effect of GMP-PNP on binding of ^{3}H-D-ala^{2}-methionine enkephalin to several rat brain regions

Brain region	IC_{50} of GMP-PNP (μM)
Cerebral cortex	105
Hypothalamus	95
Caudate nucleus	85
Medulla oblongata	55
Midbrain	51

IC_{50} is the concentration of GMP-PNP that gave 50% inhibition of ^{3}H-D-ala^{2}-methionine enkephalin binding.

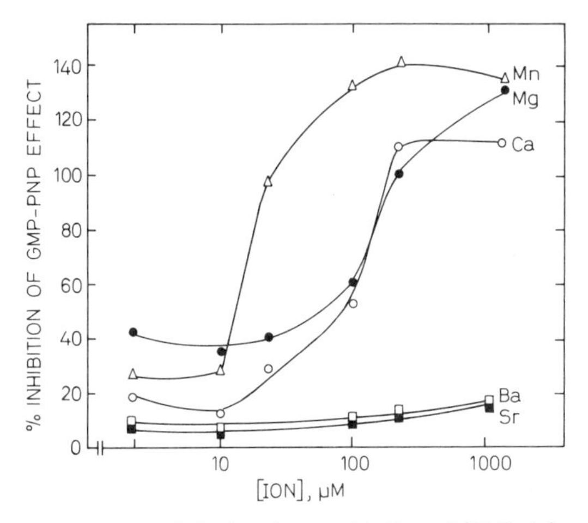

Fig. 1. Effect of divalent ions on binding of ^{3}H-D-ala^2-methionine enkephalin to opiate receptors in the presence of GMP-PNP. Crude membrane preparation of whole rat brain minus the cerebellum was prepared and the binding was conducted as described (29) in the presence of different concentrations of chloride salt of the indicated cations.

divalent ions. Calcium, magnesium and especially manganese inhibit the effect of GMP-PNP. The half maximal effect of manganese, magnesium and calcium was 10, 100 and 110 μM. These concentrations are lower than those used in the past to observe the effect of these ions on binding of opiates (3, 21, 31) or enkephalin (32). Two other divalent cations, barium and strontium, had no effect on binding of ^{3}H-D-ala^2-methionine enkephalin to opiate receptors in the presence or absence of GMP-PNP (fig. 1).

Discussion

The results of this study indicate that glucocorticoid hormones inhibit synthesis of endorphins by AtT/20 pituitary cells. Through inhibition of endorphin synthesis, there is a decrease in both the cellular content and the basal release of endorphins from these cells. This effect is not mimicked by other steroid hormones such as estradiol benzoate, protesterone, testosterone, 17-α-epitestosterone and Δ^4-androsten-11β-3,17-dione. Regulation of ACTH by glucocorticoids is thought to involve inhibition of hypothalamic ACTH releasing hormones as well as a direct inhibitory effect on pituitary cells (26 and references in it). The direct inhibitory role of glucocorticoid hormones on synthesis of endorphins by pituitary cells in culture, observed for the first time in this study, do not exclude other effects of these steroids on regulation of endorphins *in vivo*. In fact, it has been suggested (1) that dexamethasone inhibits the stimulatory effect of hypothalamic extract on release of endorphins from these cells. However, it has not been investigated whether under those conditions (1 μM dexamethasone incubated for 48 h) there was also a change in the cellular endorphin content.

The pituitary gland contains ACTH and endorphins in both the intermediate and the anterior lobes, whereas opiate receptors were detectable in the posterior lobe (33). The mechanism of regulation of ACTH and β-endorphin in the two regions of the pituitary gland that ontogenically develop from different embryonic tissues is unclear yet. It may well be that different regulatory mechanisms control the function of these peptides in the anterior and intermediate lobe of the pituitary. Indeed, it has been indicated that high concentrations of potassium stimulate release of endorphins (22, 29) and ACTH (12) from the anterior lobe but not from the intermediate lobe (13, 29). *In vivo,* adrenalectomy raised the level of endorphins and especially of enkephalin in the anterior lobe much more than in the intermediate lobe (24). Coregulation of ACTH and endorphins is further suggested by: a) the fact that both peptides coappear in the circulation after stress (10); b) both peptides share a common precursor (18); c) both are secreted from anterior pituitary cells after stimulation by hypothalamic extract or by vasopressin (1). Interestingly, acute injection of morphine into experimental animals stimulates release of ACTH and glucocorticoids. The finding that both this effect of morphine and the release of ACTH from the intermediate lobe are inhibited by dopamine agonists raises the possibility that morphine might costimulate release of ACTH and endorphins. Secretion of vasopressin is stimulated by acute morphine. Since this hormone stimulates release of endorphins from pituitary cells (1, and *A. Herz* in this volume), we suggest that acute morphine may alter endorphin secretion via activation of vasopressin release from the posterior lobe of the pituitary gland. This effect of morphine could be exerted either through the hypothalamus or directly by interaction with opiate receptors located in the posterior lobe (33). The role of glucocorticoid hormones on metabolism of endorphins under chronic opiate treatment (11) is yet unknown. However, it is worth mentioning the work initiated fifty years ago (17) and later confirmed by many studies that adrenalectomy renders rats more susceptible to the analgesic and other effects of opiates. We therefore suggest that the *in vivo* inhibitory function of glucocorticoids on the action of opiates may involve alteration of pituitary endorphins.

This study shows that GMP-PNP inhibits with different potency the binding of ^{3}H-D-ala^{2}-methionine enkephalin to opiate receptors prepared from different regions of rat brain. If the effect of this guanyl nucleotide is to increase the coupling of the membrane-bound opiate receptors to some internal components (like adenylate cyclase or guanyl nucleotide binding sites) as has been suggested for other surface-associated receptors, different sensitivities of different brain regions to GMP-PNP may reflect uneven distribution of heterogenous types of opiate receptors in rat brain. Pharmacological (19) as well as biochemical studies (16, 30) have suggested a diversity of opiate receptors in the CNS and in several peripheral organs. A further possibility is that opiate receptors in different brain regions may represent different degrees of coupling to the internal components of the cell. This is in line with studies suggesting that one type of receptor (e.g. β-adrenergic receptors) may have a wide range of coupling efficiency in different cell types.

This study also indicates that the effect of GMP-PNP on enkephalin binding is specifically blocked by the divalent ions manganese, calcium and magnesium but not by barium or strontium. Manganese was active in lower concentrations than those previously observed to affect opiates and enkephalin binding (3, 21, 32) in the absence of GMP-PNP. It is suggested that the differential effects of divalent ions on the binding of opiate agonists, antagonists and peptides may reflect alterations in the coupling of the opiate receptors to other membrane-associated constituents.

Summary

Glucocorticoids such as dexamethasone and prednisolone at physiological concentrations inhibit the synthesis, and hence the release, of endorphins by AtT/20 pituitary cells. Several other steroids, including progesterone, estradiol benzoate and testosterone, have no such effect and do not compete with dexamethasone. Incorporation of ^{3}H-dexamethasone into nuclei of AtT/20 cells was inhibited by a much lower concentration of dexamethasone or prednisolone than that of several other steroids studied. This study indicates for the first time that glucocorticoids directly inhibit synthesis of endorphins by pituitary cells.

The effect of GTP and GMP-PNP on the binding of ^{3}H-D-ala-methionine enkephalin to rat brain opiate receptors was studied. Different brain regions showed different sensitivity to GMP-PNP. Manganese, calcium and magnesium selectively inhibited the effect of GMP-PNP but barium and strontium had no effect. The possibility that divalent cations alter the coupling of opiate receptors to internal components is suggested.

References

1 Allen, R.G.; Herbert, E.; Hinman, M.; Shibuya, H., and Pert, C.B.: Coordinate control of corticotropin, β-lipotropin and β-endorphin release in mouse pituitary cell cultures. Proc. natn. Acad. Sci. USA *75:* 4972–4976 (1978).
2 Barraclough, C.A. and Sawyer, C.H.: Inhibition of the release of pituitary ovulatory hormone in the rat by morphine. Endocrinology *57:* 329–337 (1955).
3 Blume, A.J.: Interaction of ligands with opiate receptors of brain membranes: regulation by ions and nucleotides. Proc. natn. Acad. Sci. USA *75:* 1713–1717 (1978).
4 Chihara, K.; Arimura, A.; Coy, D.H., and Schally, A.V.: Studies on the interaction of endorphins, substance P and endogenous somatostatin in growth hormone and prolactin release in rats. Endocrinology *102:* 281–290 (1978).
5 Childers, S.R. and Snyder, S.H.: Guanine nucleotides differentiate agonist and antagonist interactions with opiate receptors. Life Sci. *23:* 759–762 (1978).

6 Cicero, T.J.; Meyer, E.R.; Wiest, W.G.; Olney, J.W., and Bell, R.D.: Effects of chronic morphine administration on the reproductive system of male rat. J. Pharmacol. exp. Ther. *192:* 542–548 (1975).
7 Cox, B.M.; Opheim, K.E.; Teschemacher, H., and Goldstein, A.: A peptide-like substance from pituitary that acts like morphine. Purification and properties. Life Sci. *16:* 1777–1782 (1975).
8 George, R.: Hypothalamus: anterior pituitary gland; in Clouet, Narcotic drugs: Biochemical pharmacology; pp. 283–299 (Plenum Press, New York 1971).
9 Grandison, L. and Guidotti, A.: Regulation of prolactin release by endogenous opiates. Nature, Lond. *270:* 357–359 (1977).
10 Guillemin, R.; Vargo, T.; Rossier, J.; Minick, S.; Ling, N.; Rivier, C.; Vale, W., and Bloom, F.: β-Endorphin and adrenocorticotropin are secreted concomitantly by the pituitary gland. Science, N.Y. *197:* 1367–1369 (1977).
11 Höllt, V.; Przewłocki, R., and Herz, A.: β-Endorphin-like immuno-reactivity in plasma, pituitaries and hypothalamus of rats following treatment with opiates. Life Sci. *23:* 1057–1066 (1978).
12 Kraicer, J.; Milligan, J.V.; Gosbee, J.L.; Conrad, R.G., and Branson, C.M.: *In vitro* release of ACTH: effect of potassium, calcium and corticosterone. Endocrinology *85:* 1144–1153 (1969).
13 Kraicer, J. and Morris, A.R.: *In vitro* release of ACTH from dispersed rats pars intermedia cells: effect of secretagogues. Neuroendocrinology *20:* 79–96 (1976).
14 Lazarus, L.H.; Ling, N., and Guillemin, R.: β-Lipotropin as a prohormone for the morphinomimetic peptides endorphins and enkephalins. Proc. natn. Acad. Sci. USA *73:* 2156–2159 (1976).
15 Li, C.H. and Chung, D.: Isolation and structure of an untrikontapeptide with opiate activity from camel pituitary glands. Proc. natn. Acad. Sci. USA *73:* 1145–1148 (1976).
16 Lord, J.A.H.; Waterfield, A.A.; Hughes, J., and Kosterlitz, H.W.: Endogenous opioid peptides: multiple agonists and receptors. Nature, Lond. *267:* 495–499 (1977).
17 Mackay, E.M. and Mackay, L.L.: Susceptibility of adrenalectomized rats to morphine intoxication. J. Pharmac. exp. Ther. *35:* 67–74 (1929).
18 Mains, R.E.; Eipper, B.A., and Ling, N.: Common precursor to corticotropins and endorphins. Proc. natn. Acad. Scie. USA *74:* 3014–3018 (1977).
19 Martin, W.R.; Eades, C.G.; Thompson, J.A.; Huppler, R.E., and Gilbert, P.E.: The effect of morphine and nalorphine-like drugs in the nondependent and morphine-dependent chronic spinal dog. J. Pharmac. exp. Ther. *197:* 517–532 (1976).
20 Nikodijevic, O. and Maickel, R.P.: Some effects of morphine on pituitary-adrenocortical function in the rat. Biochem. Pharmacol. *16:* 2137–2142 (1967).
21 Pasternak, G.W.; Snowman, A.M., and Snyder, S.H.: Selective enhancement of ^{3}H-opiate agonist binding by divalent cations. Mol. Pharmacol. *11:* 735–744 (1975).
22 Przewłocki, R.; Höllt, V., and Herz, A.: Release of β-endorphin from rat pituitary *in vitro*. Eur. J. Pharmacol. (in press).
23 Rennels, E.G.: Effect of morphine on pituitary cytology and gonadotropic levels in the rat. Tex. Rep. Biol. Med. *19:* 646–657 (1961).
24 Rossier, J.; Vargo, T.M.; Minick, S.; Ling, N.; Bloom, F.E., and Guillemin, R.: Regional dissociation of β-endorphin and enkephalin contents in rat brain and pituitary. Proc. natn. Acad. Sci. USA *74:* 5162–5165 (1977).
25 Rotsztejn, W.H.; Drouva, S.V.; Pattou, E., and Kordon, C.: Met-enkephalin inhibits *in vitro* dopamine induced LHRH release from mediobasal hypothalamus of male rats. Nature, Lond. *274:* 281–282 (1978).

26 Russell, S.M.; Dhariwal, A.P.S.; McCann, S.M., and Yates, F.E.: Inhibition by dexamethasone of the *in vivo* pituitary response to corticotropin-releasing factor (CRF). Endocrinology *85:* 512–521 (1969).
27 Shaar, C.J.; Frederickson, R.C.A.; Diniger, N.B., and Jackson, L.: Enkephalin analogues and naloxone modulate the release of growth hormone and prolactin: evidence for regulation by an endogenous opioid peptide in brain. Life Sci. *21:* 853–860 (1977).
28 Shuster, M.A. and Browning, B.: Morphine inhibition of plasma corticosteroid levels in chronic venous-catheterized rats. Am. J. Physiol. *200:* 1032–1034 (1961).
29 Simantov, R.: Basal and potassium stimulated calcium dependent endorphins release from pituitary cells. Life Sci. *23:* 2503–2508 (1978).
30 Simantov, R.: Enkephalins, endorphins and opiate receptors: studies *in vitro* and *in vivo*; in Graf, Palkovits and Ronai, Endorphins; pp. 221–236 (Akademia Kiado, Budapest 1978).
31 Simantov, R.; Snowman, A.M., and Snyder, S.H.: Temperature and ionic influences on opiate receptor binding. Mol. Pharmacol. *12:* 977–986 (1976).
32 Simantov, R. and Snyder, S.H.: Morphine like peptides in mammalian brain: isolation, structure elucidation and interaction with the opiate receptor. Proc. natn. Acad. Sci. USA *73:* 2515–2519 (1976).
33 Simantov, R. and Snyder, S.H.: Opiate receptor binding in the pituitary gland. Brain Res. *124:* 178–184 (1977).

Dr. R. Simantov, Department of Genetics, The Weizmann Institute of Science, Rehovot (Israel)

Prog. biochem. Pharmacol., vol. 16, pp. 32–40 (Karger, Basel 1980)

Characterization of Humoral Endorphin[1]

B.A. Weissman, R. Azov,[2] M. Granat,[3] Y. Gothilf[4] and Y. Sarne[4]

Department of Pharmacology, Israel Institute for Biological Research, Ness Ziona; [2] Department of Pharmacology, Faculty of Medicine, Technion – Israel Institute of Technology, Haifa; [3] Department of Obstetrics and Gynecology, Hadassah University Hospital, Jerusalem; and [4] Department of Physiology and Pharmacology, Sackler School of Medicine, Tel Aviv University, Tel Aviv

Introduction

Following the isolation and characterization of the pentapeptide enkephalins (15), there is an ever-growing body of information concerning endogenous opiates. In addition to the enkephalins which are mainly located in the central nervous system, there are other groups of compounds in various body fluids and tissues. β-Endorphin (3) is a higher molecular weight peptide which is concentrated in the pituitary but is also present in blood and amniotic fluid (AF) (6). These endogenous morphine-like substances exhibit opiate activity in various tests such as binding to opiate receptors, induction of analgesia (9), hormone secretion (2) and addiction (18), in a manner resembling that of the opiate alkaloids.

Other endogenous compounds, presumably of different chemical nature, have been detected in blood (11) and cerebrospinal fluid (CSF) (15, 17). Recently we reported the presence of an enkephalin-like immunoreactive substance in human CSF (13). This substance ('humoral endorphin') crossreacts with anti-leu-enkephalin antibodies although its molecular weight (1,000–1,400 daltons) is significantly higher than that of leu- or met-enkephalin.

These observations raised some questions as to the distribution and stability of humoral endorphin, as well as its biological activity and physiological role. We therefore studied its biological activity by means of an opiate receptor assay and a guinea pig ileum assay. We also measured its levels in body fluids of pregnant women.

[1] This work was partially supported by the Israel Center for Psychobiology, The Charles Smith Foundation Grants 17/79 and 177/77–18.

Materials and Methods

Samples

a. Rat brain homogenate. Rats were killed by decapitation and the brains were immediately excised and homogenized in 2 volumes of ice-cold 5% trichloroacetic acid (TCA). After centrifugation at 4°C the supernatant, containing [^{3}H] leu-enkephalin, 45.6 Ci/mmol (Amersham) to monitor the recovery, was extracted with ether to remove the acid. The lyophilized samples were reconstituted either in 0.2 M acetic acid (for column chromatography) or in 0.05 M Tris-HCl buffer (pH 7.4) containing 0.002 M EDTA (for the radioimmunoassay (RIA)). Other body fluids were treated in a similar way, except that a 100% solution of TCA was added to achieve a final concentration of 5%.

b. CSF. Human CSF was obtained from 25 patients, none of whom had a known history of narcotic medication. Samples of fluid (2–10 ml) were withdrawn from the lumbar region. In certain instances fresh fluid was used, but in most cases CSF was first treated with TCA.

c. Blood. Venous blood was taken from 40 healthy volunteers, 25–45 years old, and venous as well as cord blood was withdrawn from women during labor or cesarean sections. All blood samples were immediately centrifuged and stored at –18°C until use.

d. AF. Twelve AF samples were obtained in mid-pregnancy by transabdominal amniocentesis in the course of induced mid-trimester abortion (n = 9), or directly from the amniotic sac during inevitable premature labour (n = 3). Twenty AF samples from term pregnancies were withdrawn transvaginally under direct vision by puncturing the membrane (n = 17), or at cesarean section (n = 3).

Samples were stored at –18°C until treated with TCA.

Gel Filtration

Column chromatography was carried out as previously reported (13), using Bio Rad P-2 and Sephadex G-10 resins.

RIA

For RIA we utilized antibodies against leu-enkephalin produced in rabbits (19), as described in a recent report (13).

Bioassay

The effects of humoral endorphin on the electrically stimulated contractions of the guinea pig ileum were tested following the procedure of *Cox and Weinstock* (4). The opiate specificity of the twitch inhibition was estimated by the opiate antagonist naloxone (Endo Laboratories).

Opiate Receptor Assay

The ability of humoral endorphin to displace [^{3}H] leu-enkephalin from its binding sites in an opiate receptor assay was examined according to the method described by *Czlonkowski et al.* (5). In general, homogenates from rat striatum and frontal cortex (in 0.05 M Tris-HCl buffer (pH 7.4) containing 25 μM bacitracin and 5 nM [^{3}H] leu-enkephalin) were incubated for 15 min at 30°C. Ice-cold buffer (3 ml) was added and the mixture was rapidly filtered on GF/B filters, washed twice with 7 ml buffer and counted in a xylene:triton x-100 scintillation fluid.

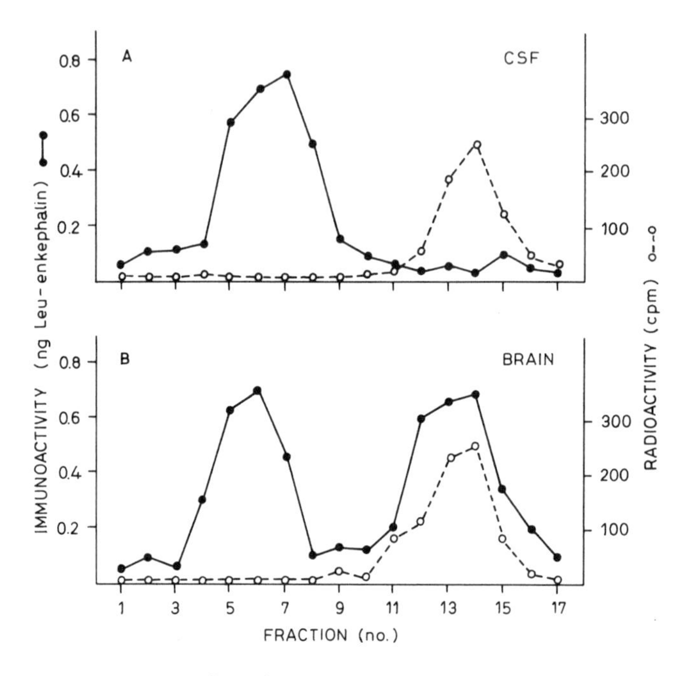

Fig. 1. Elution profile of human CSF (A) and rat brain (B) chromatographed on P-2 column. Either 0.4 ml of CSF or 0.4 ml of brain homogenate containing 0.1 g tissue was eluted with 0.2 M acetic acid on 170 × 16 mm column. Each fraction (1 ml) of the included volume (<1,800 daltons) was assayed for immunoreactive enkephalin after TCA treatment (●——●). [^{3}H] leu-enkephalin was co-chromatographed with the sample, and the radioactivity in each fraction was detected by liquid scintillation (○-----○).

Results

As can be seen in figures 1 and 2, rat brain homogenate as well as human CSF and AF contain an immunoreactive fraction which crossreacts with anti-leu-enkephalin antibodies. Human blood exhibits the same elution profile as CSF on Bio-Rad P-2 columns (results not shown). According to the P-2 chromatography, the apparent molecular weight of the immunoreactive substance is in the region of 1,000–1,400 daltons. It should be noted that rat brain homogenates but not body fluids show an elution pattern constituted from two peaks, one of which coincides with radioactive leu- or met-enkephalin (figs. 1, 2).

Humoral endorphin exhibits marked stability in human CSF, especially in comparison to leu-enkephalin. While the latter underwent 40% degradation at 21°C after 24 h incubation, the humoral endorphin level remained almost unchanged (13). At 37°C the decrease in humoral endorphin content of human CSF was less than 10%. In contrast, leu-enkephalin was rapidly degraded and

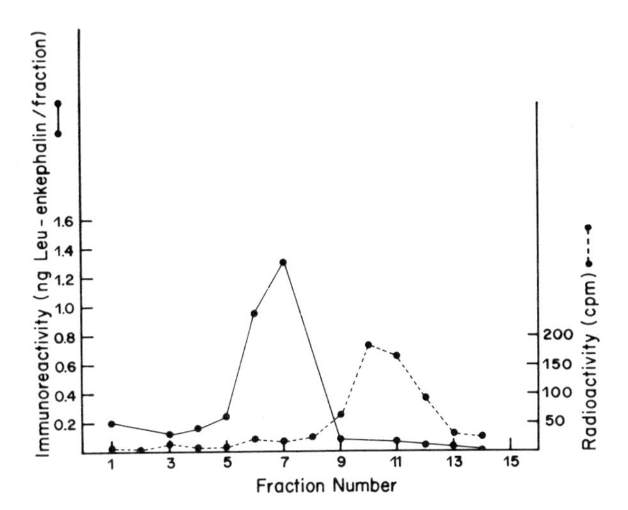

Fig. 2. Elution profile of human amniotic fluid (●——●) chromatographed on P-2 column. 0.4 ml of amniotic fluid was eluted with 0.2 M acetic acid on 170 × 16 mm column. Each fraction (1 ml) was assayed for humoral endorphin after TCA treatment. [^{3}H] leu-enkephalin was co-chromatographed with the sample and the radioactivity in each fraction was detected by liquid scintillation (●- - - -●).

only 30% of the original amount could be detected under the same conditions. Human serum has a stronger ability to decompose this endorphin and its incubation for 24 h at 37°C in serum resulted in 10–30% decrease in its concentration. It should be added that synthetic leu-enkephalin could not be detected in human serum following its addition and incubation under the same conditions.

Aliquots of fresh or TCA-treated human CSF and blood exert a considerable effect on the electrically stimulated guinea pig ileum (fig. 3). Thus, 0.2 ml of fresh CSF reduces the twitch height by 20–50%, depending on the humoral endorphin concentration in the original sample; pretreatment of the same sample with TCA increases its opiate activity by 40–60%. Similar results were obtained when the appropriate fractions of blood following column chromatography were applied to the bioassay (fig. 3). The decrease in ileum contractions caused by the TCA-soluble fraction of human blood could be related to the endorphin content as estimated by RIA. This effect on the guinea pig ileum could be blocked by 300 nM naloxone. It is worth noting that the potency of an active fraction of 0.2 ml of rat serum (following gel filtration) added to the 10 ml bath used in our bioassay equals that of both 20 nM morphine and of 80 nM leu-enkephalin, each of these concentrations representing the IC_{50}.

Figure 4 demonstrates that TCA-treated human blood, which was separated on the Sephadex G-10 column, displaced radiolabelled leu-enkephalin from its

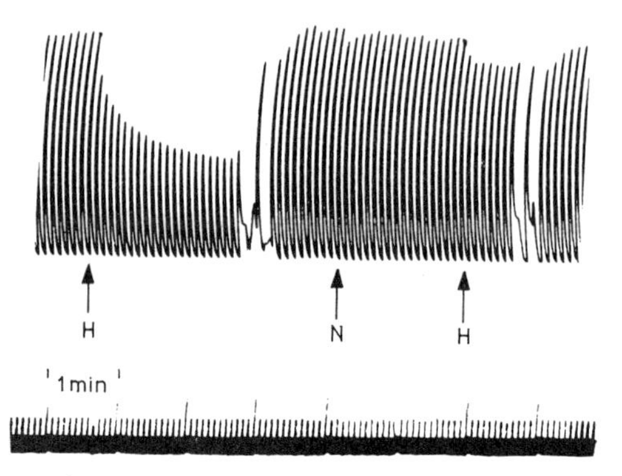

Fig. 3. The depressant effect of humoral endorphin (H, an equivalent of 0.1 ml of the TCA soluble fraction of human serum) on the electrically induced contraction of the guinea pig ileum. Naloxone (N) was added to the bath at a concentration of 300 nM and caused 90–100% reversal of the inhibition.

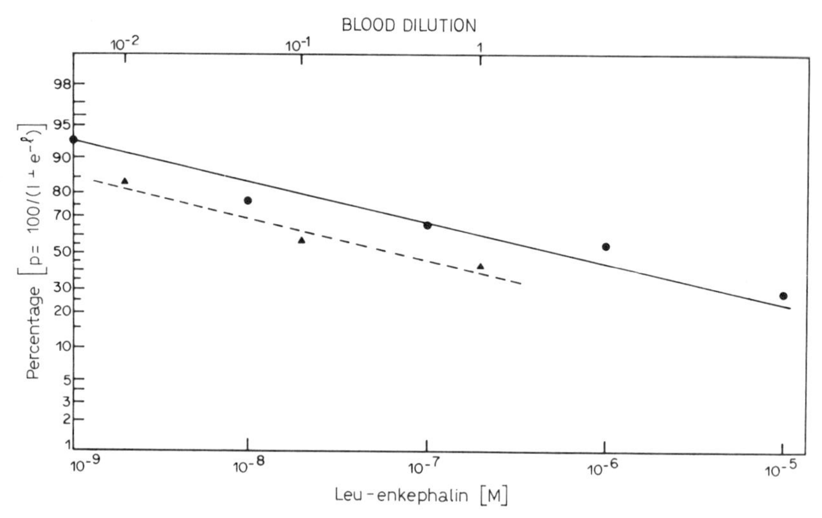

Fig. 4. Concentration dependence for inhibition of [^{3}H] leu-enkephalin binding by leu-enkephalin (●——●) and humoral endorphin (▲ – – – – ▲). Humoral endorphin was obtained from human blood (TCA soluble fraction). After Sephadex G-10 chromatography, 0.1 ml (equivalent to 0.5 ng leu-enkephalin as measured by the RIA) were introduced into the assay. Two dilutions of the original sample were also tested. The experiment was done in triplicate three times. $1 = \frac{\text{bound}}{\text{total bound}}$

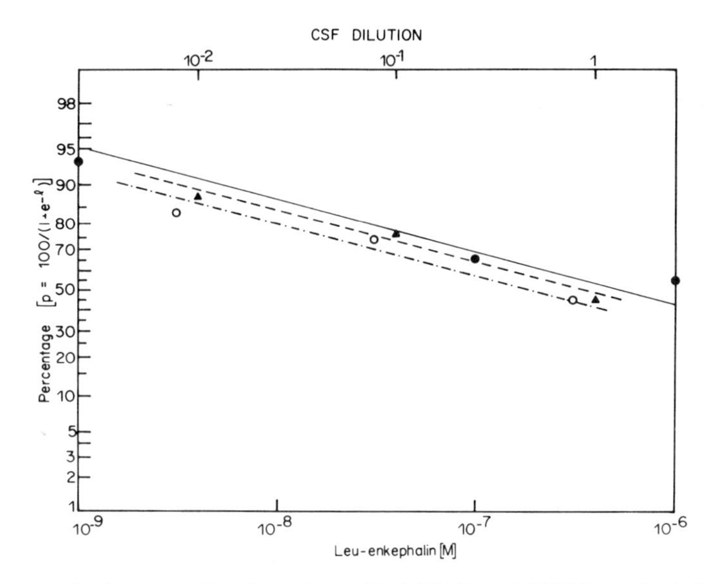

Fig. 5. Concentration dependence for inhibition of [^{3}H] leu-enkephalin binding by leu-enkephalin (•———•) and humoral endorphin. Humoral endorphin was obtained from human CSF. The original sample was divided into two parts following P-2 chromatography: A. Portion left untreated ('fresh') (▲ – – – – ▲). B. Portion treated with TCA (○– · – · –○). Two dilutions of both portions were also tested. The experiment was done in triplicate three times. $l = \frac{\text{bound}}{\text{total bound}}$

binding sites. The fact that the lines corresponding to leu-enkephalin and humoral endorphin are parallel is indicative of the competitive manner in which the latter inhibits the binding of the radioactive ligand.

Similar results were obtained when human CSF was introduced into the opiate receptor assay. After chromatography on a P-2 column, the appropriate fractions were combined and then divided into two equal portions, one of which was kept untreated ('fresh') and the other treated with TCA. It is obvious that both portions compete with leu-enkephalin on the receptor sites (fig. 5). However, TCA treatment of fractions containing humoral endorphin enhances its affinity for the opiate receptor; thus, the efficacy of the TCA-soluble fraction is 40% higher than that of the 'fresh' material.

The humoral endorphin content of AF and of maternal and cord blood during delivery is documented in table I. While AF from women delivering at mid-trimester contains 7.72 ± 0.42 ng/ml humoral endorphin (expressed as leu-enkephalin equivalents), only 5.90 ± 0.23 ng/ml could be detected in samples from women during delivery at 38–42 weeks. As can be seen from table I, there are no significant changes in the content of this substance in cord or maternal blood during pregnancy at mid-trimester or at term.

Table I. Humoral endorphin concentration in biological fluids of pregnant women[1]

Delivery type	No. of weeks	Fluid	Humoral endorphin content[2]	Range
Term	38–42	amniotic	5.90 ± 0.23 (20)*	4.52– 7.87
Early	16–28	amniotic	7.72 ± 0.42 (12)*	5.54–10.90
Term	38–42	maternal blood	14.16 ± 0.65 (22)	8.27–21.20
Early	16–28	maternal blood	15.39 ± 1.04 (4)	13.27–18.24
Term	39–42	cord blood	14.99 ± 0.99 (15)	10.94–26.25
Early	16–28	cord blood	15.20 ± 1.61 (4)	11.29–19.12

[1] Samples were obtained during delivery and treated as described in the text.
[2] Values represent $\bar{X} \pm$ S.E.M. (leu-enkephalin equivalents) with the number of samples in parentheses. * $p < 0.01$.

Discussion

Humoral endorphin is widely distributed in various body fluids and tissues of different species such as humans and rats (1, 13, 14). It should be noted, however, that while the levels of this compound are as low as 5–12 ng/ml (leu-enkephalin equivalents) in AF and CSF, blood contains 10–40 ng/ml (human) and 60–100 ng/ml (rat). β-Endorphin, which is one constituent of the heterogeneous group of endogenous opioid compounds and can also be found in all these fluids and tissues, is mainly concentrated in the pituitary (12). However, there is controversy concerning the presence of β-endorphin in the blood and CSF of normal subjects, and according to some authors this opioid can be detected only in certain pathological states (8, 10). An outstanding difference between the enkephalins and humoral endorphin is the marked resistance of the latter to degradation, as observed in blood and CSF.

The finding that humoral endorphin is biologically active in both the guinea pig ileum bioassay and the opiate receptor assay strongly suggests that it is an endogenous opiate. The ability to bind to opiate receptors in a concentration-dependent fashion and the reversal of the effect of humoral endorphin in the bioassay by naloxone are similar to the pharmacological profiles of enkephalin and β-endorphin (3, 7, 16). In contrast to the immunoreactivity of humoral endorphin which is observed only after TCA treatment, its biological activity is evident even without this treatment. It is possible that the TCA uncovers the antigenic site and thus enables the endorphin to crossreact with anti-leu-enkephalin antibodies. Moreover, some conformational changes in the molecule may occur as a result of the exposure to TCA, and this may enhance its potency in both assays.

The observation that the humoral endorphin content of AF is significantly lower than that of cord or maternal blood suggests that this substance may be of fetal origin. The fetal central nervous system as well as other fetal organs can produce this compound and release it to the amniotic sac. While there is no significant difference between humoral endorphin levels in cord and maternal blood at different gestation stages, a highly significant decrease of more than 30% is observed in the humoral endorphin content of AF in term pregnancies as compared to those at mid-trimester. We suggest that the systems responsible for metabolism and disposition of the fetus are less mature at the early stages of pregnancy and may thus allow higher rates of excretion of this endorphin to the amniotic sac.

We propose that the humoral endorphin present in blood, CSF, brain and AF may function as a neurohormone. The high concentration of humoral endorphin in the central nervous system and its presence and stability in blood and CSF provide this endogenous compound with the necessary properties to play a dual role, both as a local neurotransmitter and a systemic hormone.

Summary

Utilizing a radioimmunoassay (RIA) with antibodies produced against leu-enkephalin, the presence of humoral endorphin in various body fluids and tissues has been shown. Pretreatment with trichloroacetic acid (TCA) was needed in all cases in order to detect the immunoreactivity. However, both treated and untreated samples of humoral endorphin were active in the opiate receptor assay. Gel filtration on Bio-Gel P-2 as well as on Sephadex G-10 columns of human cerebrospinal fluid (CSF), amniotic fluid (AF) and blood shows that they all contain a similar material with an apparent molecular weight of 1,000–1,400 daltons. Chromatography of rat brain homogenate exhibited two peaks of immunoreactivity, one of which is probably due to enkephalins and the other to humoral endorphin. The latter fraction was found to be very stable when incubated in CSF, while its degradation in blood was slightly faster. This opioid compound inhibited the electrically stimulated contractions of the guinea pig ileum; the specificity of this action was indicated by its reversal with low concentrations of naloxone.

In pregnant women, humoral endorphin levels in maternal and cord blood remains stable during pregnancy, while there is a significantly higher concentration of humoral endorphin in the amniotic fluid at mid-trimester as compared to that in term pregnancies during labor.

References

1 Bergman, F.; Altstetter, R., and Weissman, B.A.: *In vivo* interaction of morphine and endogenous opiate-like peptides. Life Sci. *23:* 2601–2608 (1978).

2 Bisset, G.W.; Chowdrey, H.S., and Feldberg, W.: Release of vasopressin by enkephalin. Br. J. Pharmacol. *62:* 370–371 (1978).

3 Cox, B.M.; Opheim, K.E.; Teschemacher, H., and Goldstein, A.: A peptide-like substance from pituitary that acts like morphine. 2. Purification and properties. Life Sci. *16:* 1777–1782 (1975).

4 Cox, B.M. and Weinstock, M.: The effects of analgesic drugs on the release of acetylcholine from electrically stimulated guinea pig ileum. Br. J. Pharmacol. *27:* 81–92 (1966).

5 Czlonkowski, A.; Höllt, V., and Herz, A.: Binding of opiates and endogenous opioid peptides to neuroleptic receptor sites in the corpus striatum. Life Sci. *22:* 953–962 (1978).

6 Gantray, J.P.; Jolivet, A.; Vielh, J.P., and Guillemin, R.: Presence of immunoassayable β-endorphin in human amniotic fluid: Evaluation in cases of fetal distress. Am. J. Obstet. Gynecol. *128:* 211–221 (1977).

7 Hughes, J.; Smith, T.W.; Kosterlitz, H.W.; Fothergill, L.A.; Morgan, B.A., and Morris, H.R.: Identification of two related pentapeptides from the brain with potent opiate agonist activity. Nature, Lond. *258:* 577–579 (1975).

8 Jeffcoate, W.J.; McLoughlin, L.; Hope, J.; Rees, L.H.; Ratter, S.J.; Lowry, P.J., and Besser, G.M.: β-Endorphin in human cerebrospinal fluid. Lancet *8081:* 119–121 (1978).

9 Meglio, M.; Hosobuchi, Y.; Loh, H.H.; Adams, J.E., and Li, C.H.: β-Endorphin: Behavioral and analgesic activity in cats. Proc. natn. Acad. Sci. USA *74:* 774–776 (1977).

10 Nakai, Y.; Nakao, K.; Oki, S.; Imura, H., and Li, C.H.: Presence of immunoreactive β-endorphin in plasma of patients with Nelson's syndrome and Addison's disease. Life Sci. *23:* 2293–2298 (1978).

11 Pert, C.B.; Pert, A., and Tallman, J.F.: Isolation of a novel endogenous opiate analgesic from human blood. Proc. natn. Acad. Sci. USA *73:* 2226–2230 (1976).

12 Ross, M.; Dingledine, R.; Cox, B.M., and Goldstein, A.: Distribution of endorphins (peptides with morphine-like pharmacological activity) in pituitary. Brain Res. *124:* 523–532 (1977).

13 Sarne, Y.; Azov, R., and Weissman, B.A.: A stable enkephalin-like immunoreactive substance in human CSF. Brain Res. *151:* 399–403 (1978).

14 Shani, J.; Azov, R., and Weissman, B.A.: Enkephalin levels in rat brain after various regimens of morphine administration. Neurosci. Lett. *12:* 319–322 (1979).

15 Shorr, J.; Foley, K., and Spector, S.: Presence of a non-peptide morphine-like compound in human cerebrospinal fluid. Life Sci. *23:* 2057–2062 (1978).

16 Simantov, R. and Snyder, S.H.: Morphine-like peptides, leucine enkephalin and methionine enkephalin: Interactions with the opiate receptor. Mol. Pharmacol. *12:* 987–998 (1976).

17 Terenius, L. and Wahlström, A.: Morphine-like ligand for opiate receptors in human CSF. Life Sci. *16:* 1759–1764 (1975).

18 Wei, E. and Loh, H.: Physical dependence on opiate-like peptides. Science, N.Y. *193:* 1262–1263 (1976).

19 Weissman, B.A.; Gershon, H., and Pert, C.B.: Specific antiserum to leu-enkephalin and its use in a radioimmunoassay. FEBS Lett. *70:* 245–248 (1976).

Dr. B.A. Weissman, Department of Pharmacology,
Israel Institute for Biological Research, Ness Ziona (Israel)

Prog. biochem. Pharmacol., vol. 16, pp. 41–48 (Karger, Basel 1980)

Humoral Endorphin: Can *in vitro* Experiments Explain *in vivo* Results?[1]

Y. Sarne, Y. Gothilf and B.A. Weissman[2]

Department of Physiology and Pharmacology, Sackler School of Medicine, Tel Aviv University, Tel Aviv; and [2]Department of Pharmacology, Israel Institute for Biological Research, Ness Ziona

Introduction

The discovery of endogenous substances with opiate-like activity (2, 6, 11, 20) led to extensive research regarding the heterogeneity of the internal opiate system. The endogenous opiates (endorphins) differ from each other in their molecular structure, resulting in differences in stability and in possible physiological function. Thus, β-endorphin which is stable in blood may act as a hormone, carried to its target organs via the circulation (4), while enkephalins are degraded very rapidly (5) and may function as local neurotransmitters.

The various endorphins also differ in their distribution over the brain (16), indicating their involvement in different physiological systems. Another factor in the heterogeneity of the internal opiate system is the presence of various opiate receptors with different affinities for the various opiate agonists (9). Thus, the various endorphins may interact with different receptors, leading to differential activation of the internal system.

Recently, another endogenous opiate was detected in brain and body liquors (serum, cerebrospinal and amniotic fluids). This substance, humoral endorphin (17, 22), has some uncommon features which distinguish it from other opiates tested in the guinea pig ileum bioassay. These distinct properties of humoral endorphin may explain some puzzling results of *in vivo* experiments and may indicate its physiological function.

[1] Supported by the Israel Center for Psychobiology, The Charles Smith Foundation Grant 17/79 to *Y.S.*

Methods

Whole brains were removed immediately following decapitation of Charles River descendant rats (200–250 g) and homogenized in 5% ice-cold trichloroacetic acid (TCA). The TCA soluble fraction was freeze-dried after ether extraction of the acid. Venous serum and lumbar cerebrospinal fluid (CSF) of humans, as well as trunk serum of rats, were similarly treated with TCA.

The TCA soluble fractions of brain, CSF and serum were chromatographed on Bio Gel P-2 and Sephadex G-10 columns (40 × 4 cm) eluted with 0.2 M acetic acid. The humoral (H) endorphin fraction, detected by radioimmunoassay (17), was freeze-dried and reconstituted in KRBG solution containing NaCl (118 mM), KCl (4.75 mM), $CaCl_2$ (2.5 mM), $MgSO_4$ (1.2 mM), $NaHCO_3$ (25 mM), KH_2PO_4 (1.2 mM) and glucose (11 mM).

The opiate activity of humoral endorphin, as well as that of morphine hydrochloride (Merck, Germany), leucine-enkephalin (Miles-Yeda, Israel) and β-endorphin (Peninsula Labs, USA) was measured using the guinea pig ileum bioassay (12): An ileum strip (4–5 cm) bathed in 10 ml KRBG solution (32°C, gassed with 95% O_2–5% CO_2, pH = 7.4) was stimulated coaxially via a pair of platinum electrodes (supramaximal stimulation, 30–60 V, 0.5 msec, 0.1 or 0.15 Hz) and the isometric contraction was recorded on a polygraph.

Results

Fractions containing humoral (H) endorphin inhibited the electrically induced ileum contractions in a reversible fashion. H-endorphin extracted from 0.1 ml serum, 0.2 ml CSF or 20 mg rat brain induced 50% inhibition of contraction. The equipotent doses of morphine, β-endorphin and leucine-enkephalin in this bioassay were 0.2, 0.4 and 0.8 nmoles, respectively.

The opioid activity of H-endorphin, as well as the activity of the other opiate agonists, was blocked by 0.3 μM of naloxone, a specific opiate antagonist. However, higher concentrations of naloxone were less effective in blocking humoral endorphin, and 1 μM did not cause any reduction in the opiate activity of H-endorphin (fig. 1). Higher concentrations of naloxone even potentiated the effect of H-endorphin, in contrast to the dose-dependent antagonism of naloxone on morphine, β-endorphin and leucine-enkephalin (fig. 2).

The unexpected interaction between humoral endorphin and naloxone raised the question whether similar non-conventional interactions may take place between H-endorphin and opiate agonists. In contrast to the naloxone experiments, studying the interactions between opiate agonists provides the advantage of direct measurement of both drugs. Thus one can apply one agonist, measure its effect and follow the recovery of the preparation after removing the drug. Then another agonist is applied and its effect on the pre-treated preparation can be compared with the effect on a naive ileum. Such an experiment is shown in figure 3. Here humoral endorphin induces a constant, reproducible effect on the ileum contractions (upper trace). Morphine in a concentration which gives about the same effect (50% inhibition of contractions) is then introduced. The mor-

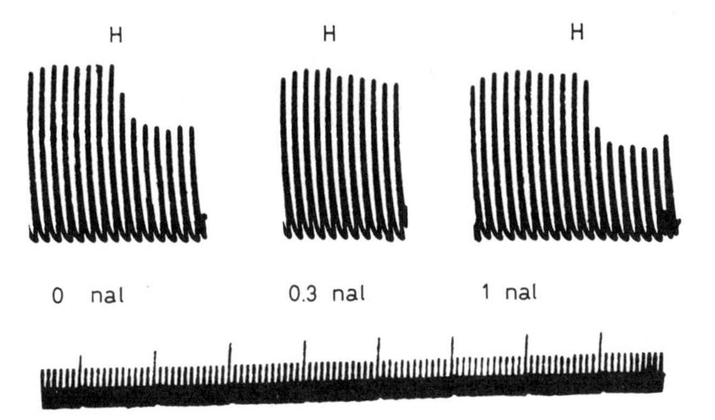

Fig. 1. Effect of naloxone on the opioid activity of humoral endorphin. Humoral endorphin (H) was applied in the absence (0 nal) and the presence of 0.3 μM and 1 μM naloxone. Time calibration: 5 sec/1 min.

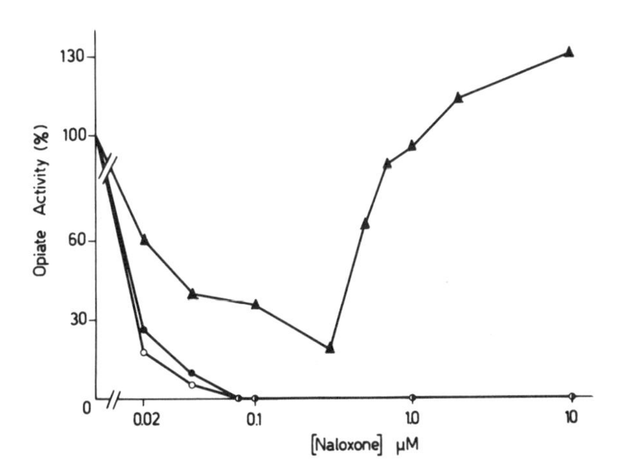

Fig. 2. Effect of naloxone on the opioid activity of humoral endorphin (▲), leucine-enkephalin (•) and morphine (○). Agonists were used in concentrations giving about 50% inhibition of the ileum twitch contraction. The maximal reversal of humoral endorphin activity was achieved by 0.2–0.4 μM naloxone, resulting in 20–100% inhibition of opiate activity in the various experiments.

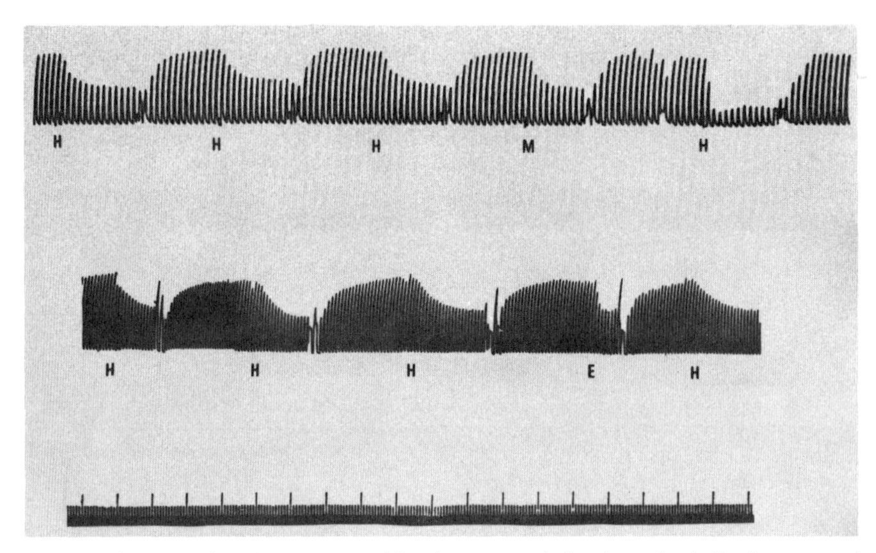

Fig. 3. Interactions between morphine (upper trace), leucine-enkephalin (lower trace) and humoral endorphin. The opiate effect of morphine (M), enkephalin (E) and humoral endorphin (H) was measured in the guinea pig ileum bioassay. Each drug was applied following washout of the previous drug and a complete recovery of the preparation. Time calibration: 5 sec/1 min.

phine is washed out and the preparation undergoes a complete recovery. However, the ileum becomes more sensitive to humoral endorphin following the pretreatment with morphine, and the same dose of H-endorphin is much more effective, causing almost 100% inhibition of contraction. This interaction between morphine and H-endorphin is specific and not common to all agonists tested in this bioassay. Figure 3 (lower trace) illustrates lack of interaction between leucine-enkephalin and humoral endorphin. Here enkephalin does not affect the responsiveness of the guinea pig ileum to H-endorphin which induces the same opioid effect as on the naive preparation. Similarly, there is no effect of morphine pretreatment on the opioid activity of leucine-enkephalin, indicating that the potentiation of H-endorphin caused by morphine is specific, and does not reflect a general sensitization of the ileum.

The interaction between morphine and humoral endorphin is reciprocal, and pretreatment of the preparation with H-endorphin potentiates the opiate effect of morphine. However, the interactions between opiate agonists are not always of a mutual nature: while morphine has no effect on the opiate activity of leucine-enkephalin (fig. 3), enkephalin potentiates the responsiveness of the guinea pig ileum to a successive application of morphine. Thus the specificity of the interactions is demonstrated not only between pairs of drugs, but also within a pair of opiate agonists.

Discussion

Some unexpected interactions between humoral endorphin and opiate agonist (morphine) and antagonist (naloxone) were observed in the guinea pig ileum bioassay. These interactions may reflect conformational changes of the opiate receptor which affect either its binding properties (affinity) or its intrinsic activity (efficacy). This transformation of the receptor was of long duration and persisted even after the drug which had induced the modulation was washed out. A similar interaction between enkephalin and morphine has been recently reported by *Vaught and Takemori* (21). It is known from other systems that drug-receptor interaction may be followed by long term modulation of the receptor. For example, acetylcholine not only activates the cholinergic receptor, but also decreases its ability to be reactivated by cholinergic agonists ('desensitization'; ref. 8) and increases its ability to bind certain partial antagonists ('metaphilic effect'; ref. 15). The opiate receptor is also accessible to modulatory effects, as sodium ions increase its affinity for naloxone and decrease its affinity for morphine (13, 19).

The differences between the various opiates with respect to their interactions with other drugs (either agonists or antagonists) could reflect the specificity of the various opiate receptors. Thus, it is possible that one type of opiate receptor is not only preferentially activated by a certain opiate (9), but also preferentially (or even exclusively) modulated by the drug. Alternatively, a common receptor which can be activated by several agonists is transformed to a more specific state which interacts with only one particular opiate.

Whatever the mechanism responsible for these interactions, the results suggest the flexibility of the internal opiate system. The question is raised whether such interactions can take place *in vivo,* resulting in functional modulation of the internal system. If morphine affects the responsiveness of the receptor to endogenous opiates, at least part of its pharmacologic action could be due to this modulatory effect rather than to direct activation of the receptor. Similarly, if endorphins modulate the responsiveness to morphine, its therapeutic effect would be dependent on the existing level of endogenous opiates. Could it be that some unexplained features of morphine reflect such synergism between endogenous and exogenous opiates? For example, the effectiveness of morphine is known to show diurnal variability (10). *Frederickson* and his co-workers (3) pointed out that the cyclic effectiveness of morphine is correlated with the cyclic activity of an internal opioid substance. If humoral endorphin facilitates the activation of opiate receptors by morphine, a dose of morphine injected while the endorphin level is high would be more effective than the same dose of morphine injected during the low-level period. Even more indicative is the interaction between humoral (H) endorphin and the opiate antagonist naloxone. The distinct property of H-endorphin, namely, lack of antagonism by high concentrations of naloxone, distinguishes it from other endorphins (β-endorphin, leucine-enkephalin) and may serve as a tool to differentiate between the various endogenous opiates *in vivo.* Several debates and paradoxical results in the literature indicate that such an interaction may take place *in vivo.* Thus, *Akil et al.* (1) were able to show that analgesia induced by brain stimulation is blocked by naloxone, suggesting that endogenous opiates mediate the analgesic effect. *Yaksh* and his co-workers (23) failed to reproduce this result using 20-fold higher doses of naloxone. Furthermore, *Pert and Walter* (14) showed that while morphine-induced analgesia is blocked by naloxone in a dose-dependent fashion, electrically induced analgesia is partially antagonized by 1 mg/kg and not affected at all by 10 mg/kg of naloxone.

Not only stimulation-induced analgesia, but also tonic perception of noxious stimuli is affected by naloxone. The increase in pain perception (hyperalgesia) induced by naloxone led *Jacob and Ramabadran* (7) to suggest tonic release of endogenous opiates which induce spontaneous analgesia. However, the hyperalgesic effect of naloxone is not monotonously dose-dependent, and high doses of the antagonist are less effective in inducing hyperalgesia (see figure 1 in ref. 7).

Another well-known effect of opiates is to stimulate hormone secretion. It was shown that the release of both growth hormone and prolactin is enhanced by morphine as well as by enkephalin analogues, and that this stimulatory effect is reversed by naloxone (18). The tonic release of these hormones is also inhibited by naloxone, suggesting that 'an enkephalinergic system may play a regulatory role in the growth hormone and prolactin release' (18). However, the effect of naloxone on the spontaneous release is distinguishable from its effect on the morphine- and enkephalin-induced release, since high doses of naloxone do not block the tonic release at all. This discrepancy between spontaneous and drug-induced release, similar to the inconsistency between spontaneous (or brain-controlled) and drug-induced analgesia, resembles the difference between humoral endorphin and all other opiates observed *in vitro* (fig. 2). This interaction with naloxone may indicate that humoral endorphin rather than enkephalin or β-endorphin is responsible for the functional regulation of pain perception and endocrine secretion.

Summary

Humoral endorphin, an endogenous substance isolated from brain, blood and cerebrospinal fluid, reveals non-conventional interactions with both opiate agonist (morphine) and antagonist (naloxone) in the guinea pig ileum bioassay. The opioid activity of humoral endorphin is potentiated by pretreatment of the preparation with morphine and *vice versa.* Naloxone, a specific opiate antagonist, interacts with humoral endorphin in a distinct manner which distinguishes it from other opiates: while low concentrations of naloxone antagonize the effect of humoral endorphin, high concentrations of the antagonist are less effective and even potentiate its opiate activity. These interactions between opiate agents can be explained assuming conformational transformation of the opiate receptor. The *in vitro* interactions shed new light on paradoxical and conflicting results of *in vivo* experiments and indicate the physiological function of humoral endorphin.

References

1 Akil, H.; Mayer, D.J., and Liebeskind, J.C.: Antagonism of stimulation-produced analgesia by naloxone, a narcotic antagonist. Science, N.Y. *191:* 961–962 (1976).

2 Cox, B.M.; Opheim, K.E.; Teschemacher, H., and Goldstein, A.: A peptide-like substance from pituitary that acts like morphine. 2. Purification and properties. Life Sci. *16:* 1777–1782 (1975).

3 Frederickson, R.C.A.; Burgis, V., and Edwards, J.D.: Hyperalgesia induced by naloxone follows diurnal rhythm in responsivity to painful stimuli. Science, N.Y. *198:* 756–758 (1977).

4 Guillemin, R.; Vargo, T.; Rossier, J.; Minick, S.; Ling, N.; Rivier, C.; Vale, W., and Bloom, F.: β-Endorphin and adrenocorticotropin are secreted concomitantly by the pituitary gland. Science, N.Y. *197:* 1367–1369 (1977).

5 Hambrook, J.H.; Morgan, B.A.; Rance, M.J., and Smith, C.F.C.: Mode of deactivation of the enkephalins by rat and human plasma and rat brain homogenates. Nature, Lond. *262:* 782–783 (1976).

6 Hughes, J.; Smith, T.W.; Kosterlitz, H.W.; Fothergill, L.A.; Morgan, B.A., and Morris, H.R.: Identification of two related pentapeptides from the brain with potent opiate agonist activity. Nature, Lond. *258:* 577–579 (1975).

7 Jacob, J.J.C. and Ramabadran, K.: Enhancement of a nociceptive reaction by opioid antagonists in mice. Br. J. Pharmacol. *64:* 91–98 (1978).

8 Katz, B. and Thesleff, S.: A study of the 'desensitization' produced by acetylcholine at the motor end-plate. J. Physiol., Lond. *138:* 63–80 (1957).

9 Lord, J.A.H.; Waterfield, A.A.; Hughes, J., and Kosterlitz, H.W.: Endogenous opioid peptides: multiple agonists and receptors. Nature, Lond. *267:* 495–499 (1977).

10 Lutsch, E.F. and Morris, R.W.: Light reversal of a morphine-induced analgesia susceptibility rhythm in mice. Experientia *27:* 420–421 (1971).

11 Pasternak, G.W.; Goodman, R., and Snyder, S.H.: An endogenous morphine-like factor in mammalian brain. Life Sci. *16:* 1765–1769 (1975).

12 Paton, D.M.: The action of morphine and related substances on contraction and acetylcholine output of coaxially stimulated guinea-pig ileum. Br. J. Pharmacol. *12:* 119–127 (1957).

13 Pert, C.B. and Snyder, S.H.: Opiate receptor binding of agonists and antagonists affected differentially by sodium. Mol. Pharmacol. *10:* 868–879 (1974).

14 Pert, A. and Walter, M.: Comparison between naloxone reversal of morphine and electrical stimulation induced analgesia in the rat mesencephalon. Life Sci. 1023–1032 (1976).

15 Rang, H.P. and Ritter, J.M.: A new kind of drug antagonism: Evidence that agonists cause a molecular change in acetylcholine receptors. Mol. Pharmacol. *5:* 394–411 (1969).

16 Rossier, J.; Vargo, T.M.; Minick, S.; Ling, N.; Bloom, F.E., and Guillemin, R.: Regional dissociation of β-endorphin and enkephalin contents in rat brain and pituitary. Proc. natn. Acad. Sci. USA *74:* 5162–5165 (1977).

17 Sarne, Y.; Azov, R., and Weissman, B.A.: A stable enkephalin-like immunoreactive substance in human CSF. Brain Res. *151:* 399–403 (1978).

18 Shaar, C.J.; Frederickson, R.C.A.; Dininger, N.B., and Jackson, L.: Enkephalin analogues and naloxone modulate the release of growth hormone and prolactin – evidence for regulation by an endogenous opioid peptide in brain. Life Sci. *21:* 853–860 (1977).

19 Simon, E.J.: The opiate receptors. Neurochem. Res. *1:* 3–28 (1976).

20 Terenius, L. and Wahlström, A.: Morphine-like ligand for opiate receptors in human CSF. Life Sci. *16:* 1759–1764 (1975).

21 Vaught, J.L. and Takemori, A.E.: Characterization of leucine and methionine enkephalin and their interaction with morphine on the guinea pig ileal longitudinal muscle. Res. Commun. chem. Path. Pharmacol. *21:* 391–407 (1978).

22 Weissman, B.A.; Azov, R.; Granat, M.; Gothilf, Y., and Sarne, Y.: Characterization of humoral endorphin. (This volume)

23 Yaksh, T.L.; Yeung, J.C., and Rudy, T.A.: An inability to antagonize with naloxone the elevated nociceptive threshold resulting from electrical stimulation of the mesencephalic central gray. Life Sci. *18:* 1193–1198 (1976).

Dr. Y. Sarne, Department of Physiology and Pharmacology,
Sackler School of Medicine, Tel Aviv University, Tel Aviv (Israel)

Prog. biochem. Pharmacol., vol. 16, pp. 49–59 (Karger, Basel 1980)

Inactivation of Enkephalin by Brain Enzymes

Z. Vogel and M. Altstein
Department of Neurobiology, The Weizmann Institute of Science, Rehovot

Introduction

There is an increasing amount of evidence suggesting that the endogenous opioid pentapeptides, the enkephalins, are putative neurotransmitters (10, 24). One of the characteristics of a neurotransmitter is its specific rapid deactivation (2). Indeed, it has been shown by several laboratories that highly effective systems for the inactivation of enkephalin exist in the brain (6, 11, 15, 17). Both Leu-enkephalin (Leu-Enk) and Met-enkephalin (Met-Enk) were shown to display a short half-life when injected into the brain, thus exhibiting only a weak and transient analgesic effect (4, 22). Enkephalin analogs with improved stability, such as (D-Ala^2)Met-Enk, (D-Ala^2)Met-Enk amide and others, have been prepared and were shown to produce more potent and longer lasting analgesic effects (19, 22). The addition of peptidase inhibitors, i.e. bacitracin or puromycin, extended the half-life of enkephalin *in vitro* (18, 23, 27). It was also shown that the intracerebral injection of bacitracin together with trasylol (inactivator of callicrein) induced a naloxone-reversible endorphin-like behavior in mice (21).

Two enzymatic mechanisms have been implicated in the inactivation of enkephalin. Incubation of Met- or Leu-Enk with brain homogenates resulted in a rapid degradation of the enkephalin molecule with the Tyr-Gly peptide bond being the preferential site of cleavage (11, 17, 26). The degradation of enkephalin by brain aminopeptidase was effectively inhibited by the antibiotic puromycin (3, 14, 27) a known inhibitor of brain arylamidase (8, 16).

In contrast to these *in vitro* studies, it has been shown by *Craves et al.* that the perfusion of labeled enkephalin through the ventricular system of rat brain resulted in the degradation of enkephalin with the coappearance of either Tyr-Gly or Tyr-Gly-Gly (6). An appropriate enzymatic activity, which releases Tyr-Gly-Gly from the enkephalin molecule, has recently been detected in mouse and rat striatum membrane preparations (15, 28). This enzymatic activity increased after chronic morphine treatment, in contrast to the aminopeptidase which was not affected (15).

In the following report we describe some of the properties of these two distinct enzymatic activities with regard to enkephalin degradation.

Materials and Methods

Materials. [Tyr-^{3}H](D-Ala^2)Met-Enk·amide was obtained from New England Nuclear. [Leu-^{3}H] Leu-Enk was synthesized by Dr. *S. Blumberg* of the Biophysics Department, Weizmann Institute of Science. Other radioactive materials were obtained from Amersham Radiochemical Centre. 5'-Deoxypuromycin (1) and the carbocyclic puromycin analog, 6-dimethylamino-9-{R-[2R-hydroxy-3R-(p-methoxyphenyl-L-alanylamino)]-cyclopentyl} purine (7) were kindly provided by Dr. *R. Vince* of the University of Minnesota, Minneapolis. The lysyl-arabinofuranosyl analog of puromycin, 9-[3'-deoxy-3'(N-L-lysylamino)-β-D arabinofuranosyl] adenine (9) was kindly donated by Dr. *S. Pestka* of the Roche Institute, Nutley, N.J. Other materials and crude brain homogenate were prepared or obtained as previously described (26, 27). Washed synaptic plasma membranes from the striata of male rats (Sprague Dawley) or male mice (C_{57}B1-6J) were purified essentially according to the flotation-sedimentation sucrose gradient centrifugation method of *Jones and Matus* (13). (See also Ref. 28.) Crude guinea pig ileum homogenate was prepared by the homogenization of the ileum in 10 volumes of 50 mM Tris-HCl, pH 7.5. The homogenate was spun for 10 min at 1,000 × g and the pellet discarded.

Enzymatic Hydrolysis of the Enkephalins. Mixtures of labeled and unlabeled enkephalin (40,000 cpm and 0.1 μM final concentration, unless otherwise indicated) were incubated with the enzymatic activity for 10 min at 30° in a final volume of 100 μl of 10 mM Tris-HCl, pH 7.5. Incubations were stopped by boiling and the reaction mixtures filtered through columns of Porapak Q to which the enkephalin but not the liberated tyrosine or Tyr-Gly-Gly is adsorbed (26). In addition, the reaction products were in some cases analyzed by thin layer chromatography (silica plates in ethyl acetate-isopropanol-acetic acid-water 40:40:1:19) or high-voltage paper electrophoresis (90 min at 3,000 V, pyridine-acetate, pH 3.5).

Binding to the Opiate Receptor. The stereospecific binding of Leu-Enk and etorphine to the receptor in crude brain homogenate was assayed as previously described (27).

Results

Hydrolysis of Enkephalin by Aminopeptidase

The major enkephalin-cleaving activity found in the crude rat brain homogenate was shown to be a neutral aminopeptidase (pH optimum of 7.0) cleaving the Tyr-Gly peptide bond of enkephalin. This enzyme was inhibited by heavy metal ions (50% inhibition with 1 μM Cd^{++} and 8 μM Zn^{++}) and by sulfhydryl blocking reagents (50% inhibition with 1 μM p-chloromercuribenzoate (PCMB), 10 μM N-ethylmaleimide and 500 μM iodoacetamide) (25). The enzymatic activity was reduced upon long exposures to air unless dithiothreitol or β-mercaptoethanol were added for stabilization. The enzyme was not sensitive to reagents which inhibit serine proteases, e.g. 0.1 mM concentrations of diethyl-

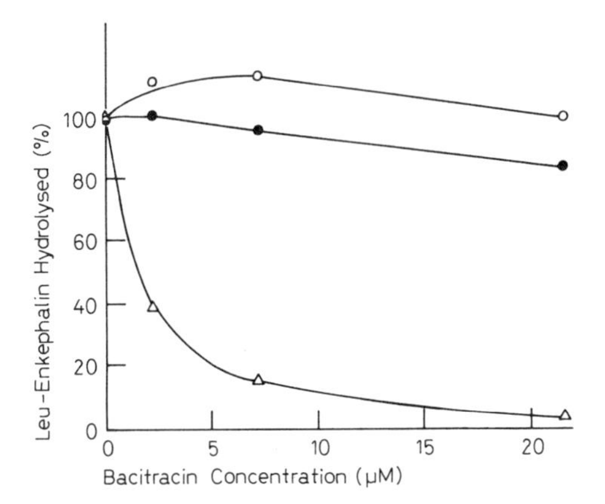

Fig. 1. The effect of bacitracin on the rate of hydrolysis of Leu-Enk by leucine aminopeptidase (○———○); rat serum (●———●); and crude rat brain homogenate (△———△). Assay conditions were as previously described (27).

fluorophosphate or phenylmethylsulfonylfluoride. Alkaloid opiates such as naloxone or morphine did not affect the enzymatic activity. The Michaelis-Menten constant for this hydrolytic activity was determined to be 2×10^{-5} M and the V_{max} was 40 nmoles/min of Leu-Enk per mg protein of crude brain homogenate at 30° (26).

The brain aminopeptidase was inhibited by the polypeptide antibiotic bacitracin (fig. 1). An inhibition of 50% was observed with 1 µM bacitracin. In contrast to the activity present in brain homogenate, the aminopeptidase activity present in rat serum as well as commercial leucine aminopeptidase were not affected by bacitracin. In accordance with this result, the binding of labeled enkephalin to the opiate receptors in crude brain homogenate was increased by the addition of bacitracin to the incubation mixture (fig. 2). On the other hand, the binding of labeled etorphine (which is not hydrolyzed by the enzyme) was not affected by low concentrations of bacitracin and was reduced somewhat with high bacitracin concentrations.

About half of the enkephalin degrading aminopeptidase was found in the soluble fraction of the homogenized brain tissue (supernatant of 20 min at 12,000 × g). The rest cosedimented with the crude mitochondrial pellet, but was largely released by hypotonic lysis of the crude synaptosomes and by further washing through a discontinuous sucrose gradient (28).

We have recently shown that puromycin is an effective inhibitor of the hydrolysis of enkephalin by the brain aminopeptidase (27). The enzymatic activity was inhibited by 50% with 0.2 µM puromycin (fig. 3). Complete inhibition (95–98%) was observed with 10 µM puromycin. As with bacitracin, puromycin

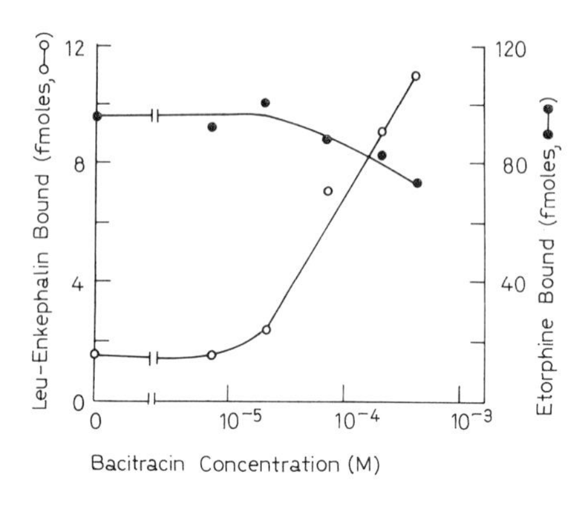

Fig. 2. The effect of bacitracin on the specific binding of etorphine or Leu-Enk to crude rat brain homogenate.

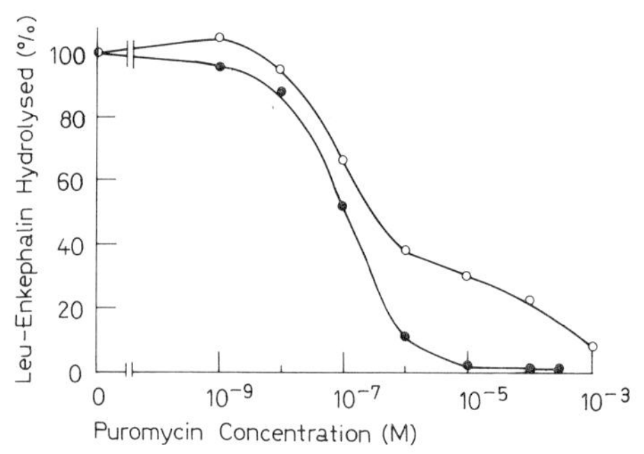

Fig. 3. The effect of puromycin on the rate of hydrolysis of Leu-Enk by 1.0 μg of crude rat brain homogenate (●——●); and by 0.65 μg of crude guinea pig ileum homogenate (○——○).

had no effect on the hydrolysis of Leu-Enk by either rat serum or by commercial leucine aminopeptidase (27). The degradation of enkephalin by homogenates of guinea pig ileum was also inhibited by puromycin, although with lower efficiency. Thus, in the presence of 10 μM puromycin the rate of hydrolysis was reduced by 70% (fig. 3).

The protection against hydrolysis served to prolong the biological activity of Leu-Enk in the guinea pig ileum preparation (fig. 4). The addition of 0.2 μM Leu-Enk to the electrically stimulated guinea pig ileum abolished the induced contractions. This effect of enkephalin was short-lived and the contractions gradually increased, approaching the original level. Puromycin alone at 0.1 μM had a very small effect, but it greatly prolonged the depressant effect of enkephalin on the contractions of the ileum.

The capacity of various puromycin analogs or derivatives to inhibit the hydrolysis of the enkephalins by brain aminopeptidase was tested. Depurination of puromycin resulted in complete loss of activity (data not shown). Similarly, the activity of puromycin was found to be dependent on the integrity of the peptide bond between the aminonucleoside and the p-methoxy-L-phenylalanine moieties of the puromycin molecule, since neither the free puromycin amino-

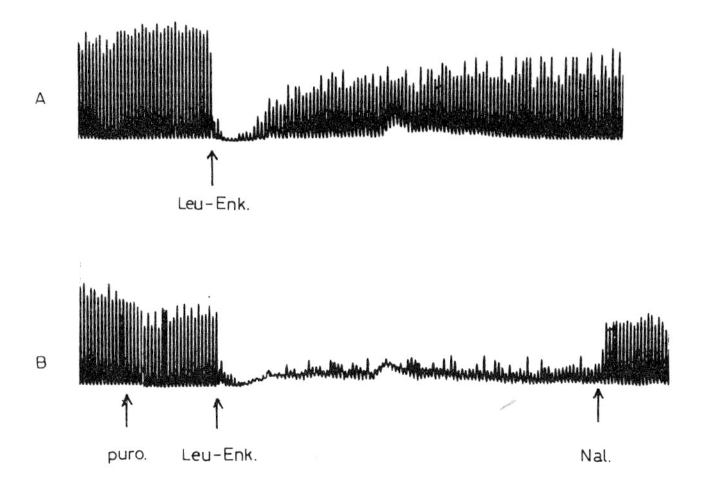

Fig. 4. The effects of puromycin and of Leu-Enk on the electrically induced contractions of the guinea pig ileum. Arrows show the addition of 0.2 μM Leu-Enk, 0.1 mM puromycin and 1 μM naloxone respectively. Figure taken from Ref. (27).

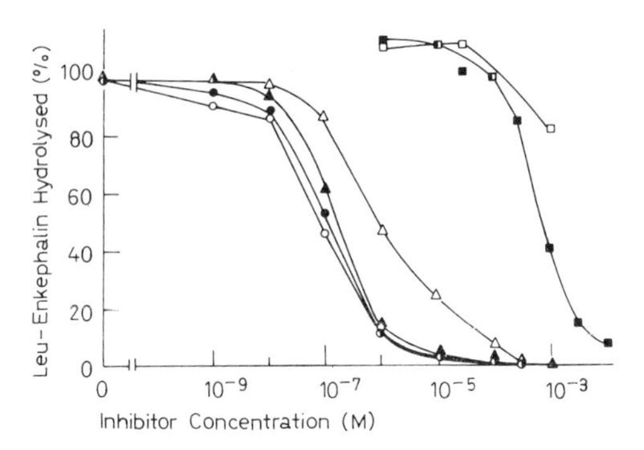

Fig. 5. The effect of increasing concentrations of puromycin and of related compounds on the rate of hydrolysis of Leu-Enk by crude rat brain homogenate. ●——●, puromycin; ○——○, 5′-deoxypuromycin; ▲——▲, lysyl-arabinofuranosyl analog of puromycin; △——△, carbocyclic analog of puromycin; ■——■, puromycin aminonucleoside; □——□, hydroxypuromycin.

nucleoside (fig. 5) nor the free p-methoxy-L-phenylalanine (see Ref. 27) were effective as inhibitors.

As is evident from figure 5, the N-terminal amino group of the p-methoxy-L-phenylalanine moiety of puromycin was also essential; hydroxypuromycin, obtained by deamination of the puromycin with nitrous acid, was ineffective as an inhibitor of enkephalin degradation. On the other hand, changes of the sugar moiety had relatively little effect. The inhibitory activities of 5′-deoxypuromycin and of the lysyl-arabinofuranoside analog of puromycin were similar to that of puromycin (50% inhibition around 0.1 μM). Similarly the carbocyclic puromycin analog, 6-dimethyl-amino-9-{R-[2R-hydroxy-3-R-(p-methoxyphenyl-L-alanylamino)]-cyclopentyl} purine in which the amino ribose moiety of puromycin was exchanged with an aminocyclopentyl group was also active as an inhibitor (50% inhibition at 1 μM).

Hydrolysis of Enkephalin by Brain Endopeptidase

Thin layer chromatography of the reaction products, obtained by the incubation of [Tyr-^{3}H] Leu-Enk with washed synaptic plasma membranes of mouse striatum, revealed the appearance of two labeled degradation products, Tyr-Gly-Gly and Tyr (fig. 6A). No Tyr-Gly or Tyr-Gly-Gly-Phe were found. In contrast to the production of Tyr, the production of Tyr-Gly-Gly was not affected by 100 μM PCMB (fig. 6A) or by 10 μM puromycin (data not shown). Thus, the two degradation products resulted from two different enzymatic activities, with the Tyr being formed by the small amount of contaminating aminopeptidase present in the washed synaptic plasma membranes (28). Figure 6B demonstrates that the Tyr-Gly-Gly was formed by an endopeptidase cleavage and did not result from the sequential action of carboxypeptidases. High-voltage paper electrophoresis of the reaction products of the hydrolysis of [Leu-^{3}H] Leu-Enk, incubated with the synaptic membranes in the presence of PCMB, demonstrated that Phe-Leu was the only labeled product formed. No labeled Leu was detected.

In contrast to the aminopeptidase, the endopeptidase is a particulate enzyme and its specific activity was increased during the purification of synaptic plasma membranes. No free soluble endopeptidase was detected throughout the purification procedure. As a result, a hundred-fold increase in the ratio between the two activities was achieved when synaptic membranes were prepared starting from the crude mitochondrial pellet of the mouse striatum (see Ref. 28). Treatment of the synaptic membranes with 1.0% of the nonionic detergent Triton X-100 released more than 75% of the endopeptidase into the high-speed supernatant without measurable loss of activity.

Figure 7 shows the Lineweaver-Burk plots of the two enkephalin-degrading activities present in the washed synaptic membranes. The K_m for the residual membrane-associated aminopeptidase was 2.6×10^{-5} M, close to the values obtained with the aminopeptidase in the soluble fractions (2.5×10^{-5} M and 2×10^{-5} M, Refs. 25 and 26 respectively). The K_m and V_{max} for the endopeptidase were determined to be 2.2×10^{-5} M and 2 nmoles/min/mg protein respectively.

The pH optimum for the endopeptidase activity was between 6.5 and 7.0 (data not shown). Interestingly, phosphate ions inhibited the hydrolysis of en-

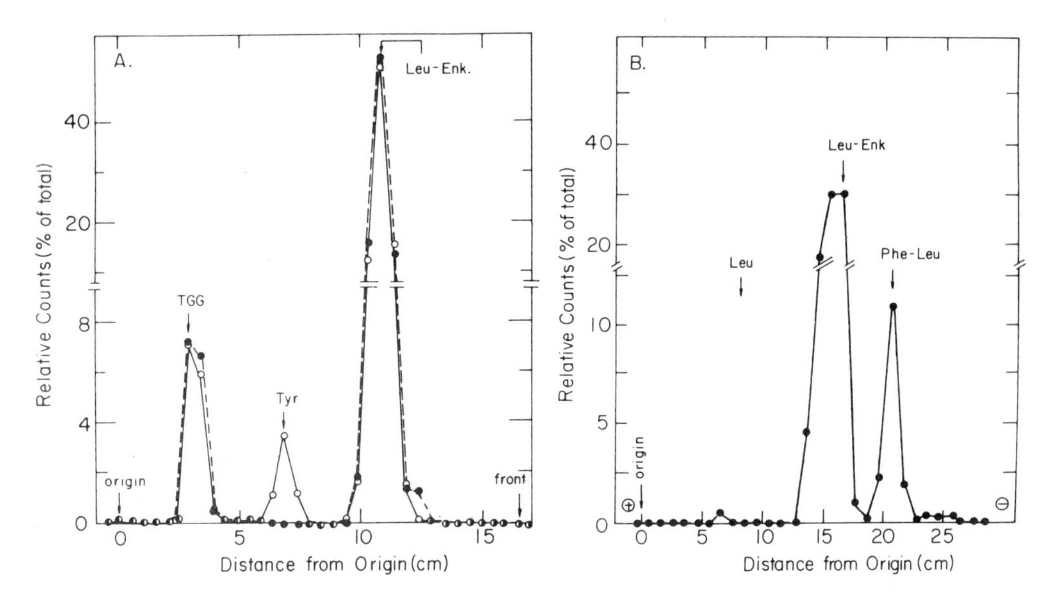

Fig. 6. A. Thin layer chromatograph of the reaction products obtained following a 20 min incubation of 50 nM [Tyr-^{3}H] Leu-Enk with 2.2 μg of mouse striatum synaptic membranes in the absence (○——○) or presence (●——●) of 0.1 mM PCMB. B. High-voltage paper electrophoresis of the reaction products obtained following 60 min incubation of 50 nM [Leu-^{3}H] Leu-Enk with 4 μg mouse striatum synaptic membranes in the presence of 0.1 mM PCMB.

kephalin by the enzyme and displaced the pH optima toward a more alkaline value. It therefore seems likely that $H_2PO_4^-$ is the inhibitory ion. The endopeptidase was inhibited by metal chelating agents, for example EGTA or 1,10-phenanthroline. It was not inhibited by 10 μM concentrations of phenylmethylsulfonylfluoride or diisopropylfluorophosphate, reagents known to inactivate proteases and other enzymes with serine in the active site. The endopeptidase was inhibited by the thiols mercaptoethanol and dithiothreitol (data not shown).

The rates of endopeptidase-induced hydrolysis of Leu-Enk, Met-Enk, and of the purportedly stable enkephalin analog (D-Ala^2)Met-Enk·amide were measured (fig. 8). Leu- and Met-enkephalin were cleaved by the rat striatum endopeptidase at approximately the same rates. (D-Ala^2)Met-Enk·amide was also hydrolyzed although at a slower rate (15%) than that of Met-Enk. Since the products of the hydrolysis of (D-Ala^2)Met-Enk·amide were not identified, it was not clear whether deamidation was a necessary step before the hydrolysis, or whether the enzyme is capable of removing the Phe-Met·amide group. The (D-Ala^2) Met-Enk·amide was completely resistant to hydrolysis by brain aminopeptidase (28).

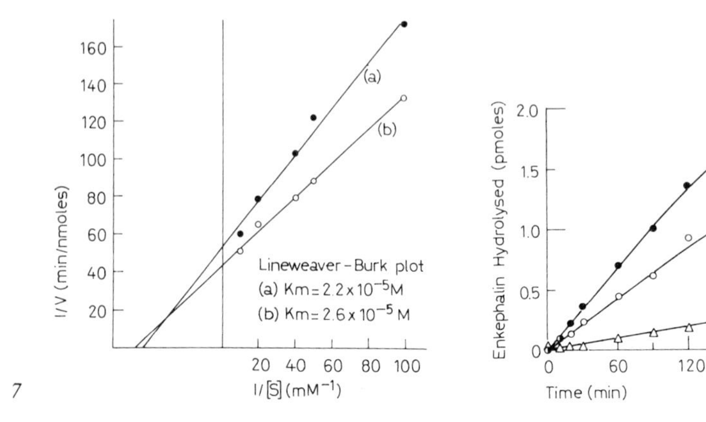

Fig. 7. Lineweaver-Burk plots of the hydrolysis of [Tyr-^{3}H] Leu-Enk by the endopeptidase (a), and by the aminopeptidase (b), present in washed synaptic plasma membranes of rat striatum. Endopeptidase activity was determined in the presence of 0.1 mM PCMB (the aminopeptidase activity was sensitive to this drug); reaction products were coanalyzed by thin layer chromatography.

Fig. 8. Time course of degradation of various labeled enkephalins by the endopeptidase present in washed synaptic plasma membranes of rat striatum. 50 nM of the labeled enkephalin were incubated for the time indicated with 4.3 μg of membrane protein in the presence of 1.0 mM PCMB. (●——●) [Tyr-^{3}H] Met-Enk; (○——○) [Tyr-^{3}H] Leu-Enk; (△——△) [Tyr-^{3}H] (D-Ala^2)Met-Enk·amide.

Discussion

Two enzymes have been implicated so far in the degradation of enkephalin in the brain. The first is a puromycin sensitive aminopeptidase (3, 14, 27), present mostly in the soluble fraction of the brain homogenate (3, 28) although some enzyme is membrane associated (14, 28). The other enzyme is a particulate, probably membrane bound, endopeptidase which cleaves enkephalin at the Gly-Phe peptide bond (15, 28).

The puromycin sensitive aminopeptidase is not restricted to the brain; similar activities have also been detected in the guinea pig ileum (5, 27), cultured neuroblastoma cells (12) and in various other cells and tissues (12, and unpublished results).

The capacity of puromycin to inhibit the hydrolysis of enkephalin by the aminopeptidase was dependent on the integrity of the glycosidic and peptide bonds connecting the structural elements of the puromycin molecule, i.e., the dimethyladenine base, the aminoribosyl moiety, and the amino acid p-methoxy-L-phenylalanine. On the other hand the p-methoxy-L-phenylalanine could be replaced by other amino acids (e.g. leucine, data not shown) and the aminoribose replaced by another aminosugar (e.g. aminoarabinofuranose) or even by the aminocyclopentyl residue, without significant reduction of the inhibitory activity. Thus, puromycin analogs could be prepared which would be ineffective as protein synthesis inhibitors (e.g. the lysyl-arabinofuranosyl analog of puromycin (20)) but still capable of inhibiting enkephalin degradation by the aminopeptidase. Such materials should find important application in the study of enkephalin degradation *in vivo.*

It is very likely that the endopeptidase described here is identical with the enzyme enkephalinase described recently by *Malfroy et al.* (15). Both enzymes are membrane associated, degrade Leu-Enk at the Gly-Phe peptide bond, and are insensitive to phenylmethylsulphonylfluoride and p-chloromercuribenzoate (or p-chloromercuriphenylsulphonate). However, there is an inconsistency in the K_m values reported. The enzyme of *Malfroy et al.* (15), obtained from the particulate fraction of mouse striatum, demonstrated a very high affinity ($K_m \cong 9 \times 10^{-8}$ M) for the hydrolysis of Leu-Enk. On the other hand our enzyme from washed mouse striatum synaptic membranes showed a K_m of 2.2×10^{-5} M (28). We analyzed the kinetic data over 6 orders of magnitude of substrate concentration (1 nM–1 mM); no other K_m value could be detected. The K_m we report is in better agreement with the known high concentration of enkephalin in the striatum and with the concept that the enkephalins are actively released and degraded in the synaptic cleft.

It is not clear whether either, both, or neither of the two enzymes, the aminopeptidase or the endopeptidase, are involved in the physiological degradation of the endogenous enkephalins. It is possible that the short half-life of the enkephalin injected into the brain (17, 22) is due to the rapid, non-specific degradation by aminopeptidase(s) (which might be released into the injection site by damage done to the tissue at the time of injection). It is likely that the relatively enhanced activity of the (D-Ala^2)-Enk analogs (19, 22) is a reflection of their stability against degradation by the puromycin sensitive aminopeptidase (28). Indeed, it was shown by the careful study of *Craves et al.*, in which enkephalin was perfused through the ventricular system of the rat brain, that the primary degradation product was not tyrosine but either Tyr-Gly or Tyr-Gly-Gly, with the free tyrosine appearing only later (6). This result is in line with the cleavage specificity of the enkephalinase/endopeptidase (15, 28). In addition, the specific activity of the enkephalinase is increased after morphine treatment (15). Moreover, the distribution of the opiate receptors shows a good correlation with that of the enkephalinase (15) but not of the aminopeptidase (*Vogel,* unpublished). These properties of the enkephalinase/endopeptidase suggest a possible role of the enzyme in the degradation of the enkephalin *in vivo.*

Summary

Two distinct enzymatic activities capable of hydrolyzing enkephalin are present in the brain. The major activity was shown to be a neutral aminopeptidase which hydrolyzes Leu-enkephalin (Leu-Enk) with a K_m of 2×10^{-5} M. This activity was inhibited by heavy metal ions (i.e. Zn^{++}, Cd^{++}), by sulfhydryl blocking reagents, and by the antibiotics bacitracin and puromycin. In contrast, these two antibiotics had no effect on the hydrolysis of Leu-Enk by either rat serum or commercial leucine aminopeptidase. The integrity of both the glycosidic and peptide bonds in the puromycin molecule was required for its inhibitory activity. On the other hand, modifications of the sugar moiety had relatively little effect, allowing the design of puromycin analogs which were inactive with regard to protein synthesis inhibition but still capable of inhibiting brain aminopeptidase. Puromycin was also shown to inhibit enkephalin degradation by homogenates of guinea pig ileum and to prolong the depressant effect of enkephalin on the electrically induced contractions of the ileum.

The second enzymatic activity in brain homogenate was found to sediment with the synaptic plasma membrane fraction and cleave Leu-enkephalin into Tyr-Gly-Gly and Phe-Leu with a K_m of 2.2×10^{-5} M and pH optimum between 6.5 and 7.0. This endopeptidase was inhibited by metal chelating agents and by thiols but was insensitive to puromycin and to p-chloromercuribenzoate. In contrast to the aminopeptidase some cleavage of (D-Ala^2)Met-Enk·amide by the endopeptidase was observed.

References

1 Almquist, R.G. and Vince, R.: Puromycin analogs; Synthesis and biological activity of 5′-deoxypuromycin and its aminonucleoside, 6-dimethylamino-9-(3′-amino-3′,5′-dideoxy-β-D-ribofuranosyl)purine. J. med. Chem. *16:* 1396–1399 (1973).

2 Aprison, M.H. and Werman, R.: A combined neurochemical and neurophysiological approach to identification of central nervous system transmitters. Neurosci. Res. *1:* 143–174 (1968).

3 Barclay, R.K. and Phillips, M.A.: Inhibition of the enzymatic degradation of Leu-enkephalin by puromycin. Biochem. biophys. Res. Commun. *81:* 1119–1123 (1978).

4 Belluzi, J.D.; Grant, N.; Garsky, V.; Sarantakis, D.; Wise, C.D., and Stein, L.: Analgesia induced *in vivo* by central administration of enkephalin in rat. Nature, Lond. *260:* 625–626 (1976).

5 Carviso, G.L. and Musacchio, J.M.: Inhibition of enkephalin degradation in the guinea pig ileum. Life Sci. *23:* 2019–2030 (1978).

6 Craves, F.B.; Law, P.Y.; Hunt, C.A., and Loh, H.H.: The metabolic disposition of radiolabeled enkephalins *in vitro* and *in situ.* J. Pharmac. exp. Ther. *206:* 492–506 (1978).

7 Daluge, S. and Vince, R.: Synthesis and antimicrobial activity of a carbocyclic puromycin analog; 6-Dimethylamino-9-{R-[2R-hydroxy-3R-(p-methoxyphenyl-L-alanylamino)]cyclopentyl}purine. J. med. Chem. *15:* 171–177 (1972).

8 Ellis, S. and Perry, M.: Pituitary arylamidases and peptidases. J. biol. Chem. *241:* 3679–3686 (1966).

9 Fisher, L.V.; Lee, W.W., and Goodman, L.: Puromycin analogs: Aminoacyl derivatives of 9-(3′-amino-3′-deoxy-β-D-arabinofuranosyl)adenine. J. med. Chem. *13:* 775–777 (1970).

10 Frederickson, R.C.A.: Enkephalin pentapeptides – a review of current evidence for a physiological role in vertebrate neurotransmission. Life Sci. *21:* 23–42 (1977).

11 Hambrook, J.M.; Morgan, B.A.; Rance, M.J., and Smith, C.F.C.: Mode of deactivation of the enkephalins by rat and human plasma and rat brain homogenates. Nature, Lond. *262:* 782–783 (1976).
12 Hazum, E.; Chang, K.J., and Cuatrecasas, P.: Rapid degradation of [^{3}H]leucine-enkephalin by intact neuroblastoma cells. Life Sci. *24:* 137–144 (1979).
13 Jones, D.H. and Matus, A.I.: Isolation of synaptic plasma membrane from brain by combined flotation-sedimentation density gradient centrifugation. Biochim. biophys. Acta *356:* 276–287 (1974).
14 Knight, M. and Klee, W.A.: The relationship between enkephalin degradation and opiate receptor occupancy. J. biol. Chem. *253:* 3843–3847 (1978).
15 Malfroy, B.; Swerts, J.P.; Guyon, A.; Rocques, B.P., and Schwartz, J.C.: High affinity enkephalin degrading peptidase in brain is increased after morphine. Nature, Lond. *276:* 523–526 (1978).
16 Marks, N.; Datta, R.K., and Lajtha, A.: Partial resolution of brain arylamidases and aminopeptidases. J. biol. Chem. *243:* 2882–2889 (1968).
17 Meek, J.L.; Yang, H.Y.T., and Costa, E.: Enkephalin catabolism *in vitro* and *in vivo*. Neuropharmacology *16:* 151–154 (1977).
18 Miller, R.J.; Chang, K.J., and Cuatrecasas, P.: The metabolic stability of the enkephalins. Biochem. biophys. Res. Commun. *74:* 1311–1317 (1977).
19 Pert, C.B.; Pert, A.; Chang, J.K., and Fong, B.T.W.: [D-Ala^2]Met-enkephalin amide: A potent, long lasting synthetic pentapeptide analgesic. Science, N.Y. *194:* 330–332 (1976).
20 Pestka, S.; Rosenfeld, H.; Harris, R., and Hintikka, H.: Studies on transfer ribonucleic acid-ribosome complexes: Effect of antibiotics on peptidyl-puromycin synthesis by mammalian polyribosomes. J. biol. Chem. *247:* 6895–6900 (1972).
21 Pinsky, C.; Brockhausen, E.; Dua, A.K.; Havlicek, V., and Labella, F.S.: Endorphin-like behavior provoked by mild stress in the presence of intracerebroventricular peptidase inhibitors; in Way, Endogenous and exogenous opiate agonists and antagonists (Pergamon Press, New York) (in press).
22 Roemer, D.; Buescher, H.H.; Hill, R.C.; Pless, J.; Bauer, W.; Cardinaux, F.; Closse, A.; Hauser, D., and Huguenin, R.: A synthetic enkephalin analogue with prolonged parenteral and oral analgesic activity. Nature, Lond. *268:* 547–549 (1977).
23 Simantov, R. and Snyder, S.H.: Morphine like peptides, leucine enkephalin and methionine enkephalin, interaction with the opiate receptor. Mol. Pharmacol. *12:* 987–998 (1976).
24 Snyder, S.H. and Childers, S.R.: Opiate receptors and opioid peptides. Ann. Rev. Neurosci. *2:* 35–64 (1979).
25 Vogel, Z. and Altstein, M.: Enzymatic cleavage of the opioid peptide leucine enkephalin. Israel J. med. Sci. *13:* 943 (1977).
26 Vogel, Z. and Altstein, M.: The adsorption of enkephalin to porous polystyrene beads, a simple assay for enkephalin hydrolysis. FEBS Lett. *80:* 332–336 (1977).
27 Vogel, Z. and Altstein, M.: The effect of puromycin on the biological activity of Leu-enkephalin. FEBS Lett. *98:* 44–48 (1979).
28 Vogel, Z. and Altstein, M.: Degradation of enkephalin by two brain enzymatic activities; in Way, Endogenous and exogenous opiate agonists and antagonists (Pergamon Press, New York) (in press).

Dr. Z. Vogel, Department of Neurobiology, The Weizmann Institute of Science, Rehovot (Israel)

Prog. biochem. Pharmacol., vol. 16, pp. 60–68 (Karger, Basel 1980)

Biosynthesis of Neuronal Peptides: Implications for Neurobiology

H. Gainer, Y.P. Loh and J.T. Russell

Laboratory of Developmental Neurobiology, National Institute of Child Health and Human Development, Bethesda, MD

Introduction

Much of our understanding of peptide biosynthesis in general comes from studies on tissues other than brain. To date, the mechanisms for biosynthesis of neuronal peptides appear to be similar to those found in other eukaryotic tissues. Two major alternatives exist: 1) the synthesis of oligopeptides by enzymatic mechanisms, i.e., synthetases; and 2) synthesis by conventional ribosomal protein synthesis mechanisms, usually as a 'prohormone', which is then degraded by limited proteolysis to specific peptide products in the cell before release. The former mechanism operates in the case of various small peptides such as carnosine (β-alanyl-L-histidine) and glutathione (γ-L-glutamyl-L-cysteinylglycine) which are found in various eukaryotic tissues (including brain) and are synthesized by enzymes such as carnosine synthetase and γ-glutamyl-L-cysteine (plus tripeptide) synthetase, respectively. Larger peptides such as insulin, nerve growth factor, ACTH, endorphin, vasopressin and oxytocin appear to be synthesized as prohormones.

A number of approaches have evolved in recent years for the study of peptide biosynthesis. At the outset it is important to determine which of the above biosynthetic mechanisms is relevant. If various inhibitors of eukaryotic protein synthesis (e.g. cycloheximide, puromycin) inhibit the synthesis of the peptide, then there is presumptive evidence for the prohormone mode of synthesis. If this is not the case, then a search for a specific enzymatic mechanism is in order. Another clue can be provided by radioimmunoassays of the peptide in order to evaluate whether, on biochemical separation (e.g., gel filtration), higher molecular weight, heterogeneous immunoreactive forms of the peptide can be detected. This also represents presumptive evidence but not proof of a prohormone, and 'big' forms of insulin, ACTH, growth hormone, calcitonin, etc. have been detected by this method. The key issue is to demonstrate biosynthesis of the peptide (and prohormone) from isotopically labeled amino acids in a pulse-chase labeling paradigm. The minimum requirement in these experiments is to show that the higher molecular weight precursor (or prohormone) decreases in radioactivity with time of the chase concurrently with an increase in the radioactivity of the peptide product. In order to demonstrate this, it is necessary in many cases (particularly with heterogeneous tissue such as brain) to resolve the relevant labeled molecules in a quantitative immunoprecipitation procedure, using an antibody to the peptide which also reacts with the precursor. Labeled presumptive precursors can be purified from the immunoprecipitates and examined for the presence of the peptide sequence by limited proteolysis mapping *in vitro* and by analysis of amino acid sequences. Finally, the use of cell-free protein synthesizing systems (e.g., wheat germ and reticulocyte *in vitro* systems) together with polyribosomes from the tissue under study can lead to the synthesis of the prohormone, *in vitro*. This approach has the added advantages not only of identifying the pre-prohormone, i.e., the prohormone with a characteristic peptide still attached to its N-terminus which is used as a signal to direct the prohormone into the cisternae of the endoplasmic reticulum, but also of representing the first step in the isolation of the m-RNA and ultimately the gene for the prohormone.

Development of the Prohormone Concept

The discovery in 1967 by *Steiner* and his colleagues that insulin is synthesized in a larger precursor form as proinsulin (33) provided the impetus for further studies which have shown that this is a common mode of biosynthesis of eukaryotic peptides destined for secretion. The continuing studies on proinsulin offer a useful intellectual and experimental paradigm (31–33). Proinsulin is a single polypeptide chain ordered as follows: NH_2-B-chain-Arg-Arg-C-peptide-Lys-Arg-A-chain-COOH. The presumed function of the C-peptide is to ensure the correct folding and sulfhydryl oxidation between the A and B chains in proinsulin. The higher rate (15 times) of mutation in the C-peptide as compared to the insulin moiety suggests that the C-peptide may not have a physiological function; on the other hand, immunoassays of C-peptide indicate that this may be a useful differential diagnostic procedure for hypoglycemia (26). The transformation of proinsulin to insulin in the beta cells begins in the Golgi apparatus where new secretory granules are formed, and continues in the granules, the half-time conversion being about 1 hour. Two major processing enzymes have been identified in the granules, a trypsin-like endopeptidase and a carboxypeptidase-B-like exopeptidase. The insulin is maintained in the granule in a crystalline form (in combination with zinc). These studies resulted in several important generalizations: a) Peptide hormones can be synthesized in larger precursor forms (i.e., as prohormones); b) Post-translational processing of the prohormone occurs in the secretory granule; and c) Peptides other than the known biologically active one(s) will emerge from this biosynthetic mechanism, and, if processing occurs intragranularly, all will be released simultaneously by the cell.

The intracellular organization involved in the biosynthesis of proteins destined for secretion has been known for some time (22). More recently, it has been discovered that the intracellular compartmentation following synthesis of the protein on the rough endoplasmic reticulum (RER) is determined, in part, by the translational process itself (5, 6, 14). The initial N-terminal amino acid sequence of the protein serves as a signal for the protein to traverse the RER membrane and enter the cisternae where the 'signal sequence' is immediately cleaved off (5, 6, 14). Similar 'signal sequences' exist for secreted proteins (5, 6), and intrinsic membrane proteins (14). In the case of prohormones, the prohormone plus its transient N-terminal sequence (usually containing 20–30 amino acids) is referred to as a 'pre-prohormone' (for examples see ref. 13).

The enzymatic mechanisms involved in the conversion of protein precursors (prohormones) to peptides (hormones) are poorly understood. In virtually all cases (13) the enzyme activities appear to be similar to those in the insulin model described above. The conversions appear to be due to the combined action of tryptic-like endopeptidases and carboxypeptidase-B-like exopeptidases. Analysis of the primary structures of various prohormones (13) indicates that they contain sequences of 2–3 basic amino acids (lysine or arginine) at the cleavage sites, and would therefore be susceptible to appropriate cleavage by the above enzymes. In all cases, pancreatic trypsin appears to mimic the endogenous protease activity. However, in no case has the enzyme activity been inhibited by specific inhibitors of pancreatic trypsin (e.g., soybean trypsin inhibitor or TLCK). It is not known whether the tryptic-like enzyme is a specific enzyme in each system.

Recently, a precursor form of β-nerve growth factor (β-NGF) has been detected in pulse-chase biosynthesis experiments (1). The precursor has an apparent molecular weight of 22,000, and is subsequently converted to the biologically active form of β-NGF (MW 13,260) by a trypsin-like enzyme. What is especially interesting is that the biologically active β-NGF is extracted from mouse salivary gland (where it is found in high concentration) in a stable 7S complex (18). The 7S NGF complex has a molecular weight of 130,000, and is composed of dimers of each of three subunits; β-NGF (MW 13,260), an α-subunit (MW 26,500), and a γ-subunit (MW 26,000). The γ-subunit is believed to be the arginine-specific esteropeptidase which converts the precursor to β-NGF (1, 18). It has been suggested that the significance of the continued binding of the enzyme (or γ-subunit) to the product (in the 7S complex) after cleavage of the precursor is related to its possible role as a mechanism for preventing further degradation by other intracellular proteases (1, 18). That this 7S complex enzyme is not 'specific' for the cleavage of β-NGF precursor has been shown in its ability to promote fibrinolysis by converting plasminogen to plasmin (21).

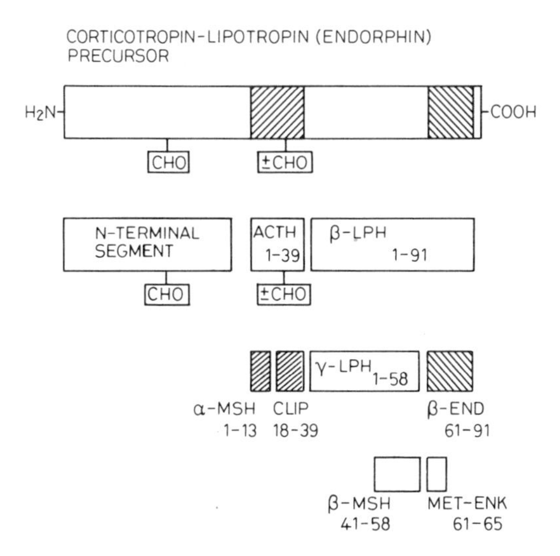

Fig. 1. Hypothetical structure of the corticotropin (ACTH)-lipotropin (endorphin) precursor, and its theoretically possible biologically active peptide products. Based on evidence given in refs. 8 and 19.

The Corticotropin-Endorphin Precursor

The existence of large immunoreactive forms of corticotropin (ACTH) has been known for many years (35). In addition, it was known that the sequences of β-MSH and α-LPH were contained in β-LPH, and that α-MSH was a tridecapeptide identical to the N-terminal sequence of ACTH (29). Because of these and other similar observations it was suspected that ACTH and MSH were formed from a larger precursor, and that β-MSH and endorphin were derived from β-LPH.

A major advance in this field was recently made as a result of studies on an ACTH-secreting mouse pituitary tumor cell line, which demonstrated that there was a common precursor of 31,000 MW for both corticotropin and β-lipotropin (8, 10, 17, 25). These elegant studies involved the use of quantitative pulse-chase experiments, immunoprecipitation by specific ACTH and endorphin antisera, and peptide mapping procedures following limited proteolysis. In addition, the common precursor has been shown to be a glycoprotein, with one of the two carbohydrate chains potentially attached to the carboxyl-terminal half of the $ACTH_{1-39}$ peptide (7). The hypothetical structure of this precursor and the potential peptide products generated from it are illustrated in figure 1. Note that there are at least six known biologically active peptides (ACTH, α-MSH, β-LPH, β-MSH, β-endorphin, and methionine enkephalin) contained within the precursor's sequence. Evidence exists for the biological relevance of this biosynthetic pathway for all but β-MSH and methionine-enkephalin (8, 9, 15). Note also that, in addition to yielding ACTH-like and endorphin-like peptides, the processing of the 31,000 MW precursor also produces a large (11,200 dalton) N-terminal glycopeptide fragment (8) of unknown function, quite reminiscent of the C-peptide and proinsulin.

When m-RNA which encodes for the common precursor is translated in a cell-free protein-synthesizing system, an unglycosylated form (MW 28,500) of the precursor is synthesized (20, 24, 25). The carbohydrate which is attached to the precursor post-translationally in the RER and Golgi appears to be important in regulating its subsequent intracellular processing (16). Recently, cDNA synthesized from the m-RNA that encodes for the common precursor in the bovine neurointermediate lobe was inserted into bacterial plasmids and cloned (19). Sequence analysis of the cDNA demonstrated that it contained the entire coding sequence for both ACTH and β-LPH, and that these two sequences are separated by a 6-base pair sequence encoding lysine and arginine (19). Since the carboxyl terminus of the ACTH in the precursor is linked via two basic amino acids to the amino terminus, it is likely that the converting enzymes will be found to be similar to the trypsin-like and carboxypeptidase-B-like activities described above for other prohormones.

Peptide Biosynthesis in Neurons

It is not commonly known that the first hypothesis for the existence of a prohormone derived from studies on the nervous system. In 1964, *Sachs and Takabatake* (28) hypothesized that vasopressin (a nonapeptide) and neurophysin (a protein of MW ~10,000, see review of neurophysins in ref. 23) were formed by post-translational processing of a common precursor protein. Evidence for this precursor was indirect (27), and only recently two precursors of neurophysin (each of MW ~20,000), one associated with vasopressin synthesis and the other with oxytocin synthesis, have been identified in pulse-chase and immunoprecipitation experiments on the rat hypothalamo-neurohypophysial system (2, 3, 11, 12). Direct evidence has recently been obtained from peptide mapping (limited proteolysis) studies that these neurophysin precursors also contain the respective nonapeptides (i.e., oxytocin and vasopressin) and are, in fact, common precursors (*Russell, Brownstein and Gainer,* in preparation). Since the precursor has a MW of 20,000 and the MW of the known peptide products sum to only about 11,000, it is expected that, as in the cases of proinsulin and the corticotropin-endorphin precursor, physiologically 'extraneous' peptides may be generated during post-translational processing of the precursors. The nature of these peptides is unknown at present.

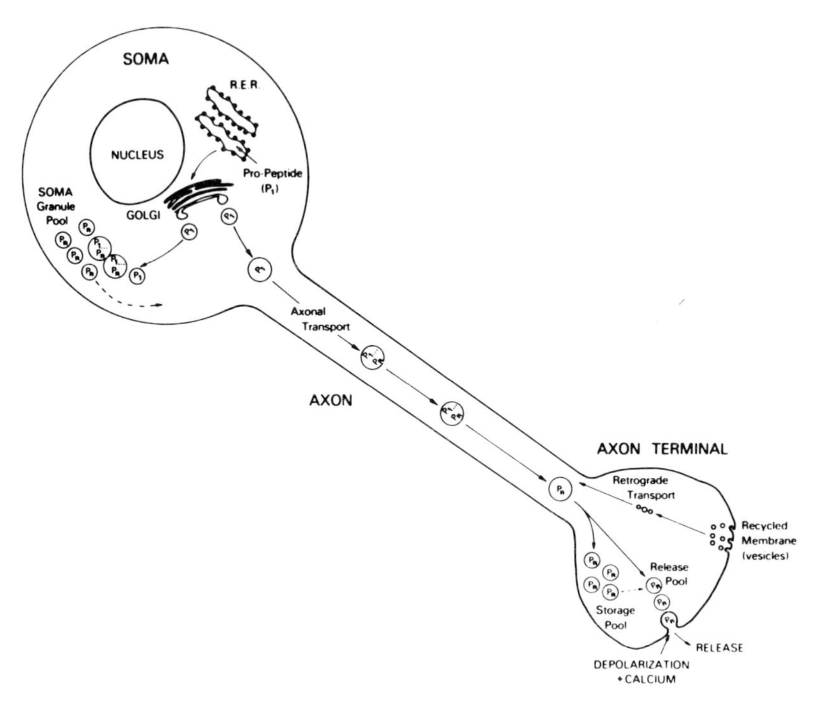

Fig. 2. Hypothetical model of biosynthesis, translocation, processing, and release of peptides in a peptidergic neuron. RER, rough endoplasmic reticulum; P_1, propeptide or precursor molecule; $P_1 \cdots P_n$, intermediates between P_1 and P_n; P_n, final peptide products of processing. From ref. 11.

The conversions of these precursors appear to occur in the secretory granules during axonal transport (11, 12). Given this site of conversion, one would expect that the unknown peptides would also be released during the stimulus-induced exocytosis of the neurophysins and their respective nonapeptides. A hypothetical model of a peptidergic neuron is illustrated in figure 2.

Implications for Neurobiology

The precursor model of peptide biosynthesis has a number of implications for the study of neuropeptide biosynthesis. First, regulation of the biosynthesis of neuronal peptides would have to be at the perikaryon level, since axons and terminals do not have ribosomes or protein synthesis machinery. Some obvious candidates for regulatory signals are the synaptic inputs to the cells, electrical activity of the neurons, and feedback of information from the release sites, possibly conveyed by retrograde axonal transport. Furthermore, the model implies that the post-translational conversions occur within the secretory granules. Hence, the processing machinery (e.g., the proteases) must be intragranular, and all the peptides generated by this process would therefore be accessible to the release (exocytosis) mechanism.

There are at least three factors determining which peptides will be released by a peptidergic neuron: a) the transcription and translation of the specific gene-encoded messages for the precursor in a given cell; b) the introduction of specific post-translational modifications in the precursor (e.g., glycosylation, phosphorylation, methylation, etc.) which may regulate its limited proteolysis; and c) the spectrum of specific proteases found in the processing site (e.g., intragranularly). Probably the best example of such regulation diversity can be found in the varied processing of the corticotropin-endorphin precursor depending on glycosylation, developmental state, and cell type (9, 16, 30).

In recent years, application of the techniques of immunocytochemistry and radioimmunoassay has provided evidence that there are projections of neurons containing peptides to various regions of the brain. One example of this is the unexpected finding that the oxytocinergic and vasopressinergic neurons in the paraventricular nucleus have widespread projections to the limbic system, medulla oblongata, and spinal cord (4, 34). In view of the biosynthetic processes described above, it seems possible that peptides other than the ones being assayed immunologically might be used as intercellular messengers in these new pathways. The study of neuropeptide biosynthesis provides an approach to the identification of these peptides.

Summary

Many biologically active peptides (e.g., insulin, nerve growth factor, ACTH, endorphin, parathyroid hormone, etc.) appear to be synthesized first as prohormones, which are then converted intracellularly to the biologically active products by various post-translational modifications. Peptides of neuronal origin (e.g., vasopressin and oxytocin) are synthesized by similar mechanisms. The prominent role of post-translational processing in determining the final peptide products allows for the possibility that different peptides will be generated from identical prohormones in different cells.

References

1 Berger, E.A. and Shooter, E.M.: Evidence for pro-β-nerve growth factor, a biosynthetic precursor to β-nerve growth factor. Proc. natn. Acad. Sci. USA *74:* 3647–3651 (1977).

2 Brownstein, M.J. and Gainer, H.: Neurophysin biosynthesis in normal rats and in rats with hereditary diabetes insipidus. Proc. natn. Acad. Sci. USA *74:* 4046–4049 (1977).

3 Brownstein, M.J.; Robinson, A.G., and Gainer, H.: Immunological identification of rat neurophysin precursors. Nature, Lond. *269:* 259–261 (1977).
4 Buijs, R.M.: Intra- and extrahypothalamic vasopressin and oxytocin pathways in the rat. Cell Tissue Res. *192:* 423–435 (1978).
5 Campbell, P.N. and Blobel, G.: The role of organelles in the chemical modification of the primary translation products of secretory proteins. FEBS Lett. *72:* 215–226 (1976).
6 Devillers-Thiery, A.; Kindt, T.; Scheele, G., and Blobel, G.: Homology in amino-terminal sequence of precursors to pancreatic secretory proteins. Proc. natn. Acad. Sci. USA *72:* 5016–5020 (1975).
7 Eipper, B.A. and Mains, R.E.: Peptide analysis of a glycoprotein form of adrenocorticotropic hormone. J. biol. Chem. *252:* 8821–8832 (1977).
8 Eipper, B.A. and Mains, R.E.: Analysis of the common precursor to corticotropin and endorphin. J. biol. Chem. *253:* 5732–5744 (1978).
9 Eipper, B.A. and Mains, R.E.: Existence of a common precursor to ACTH and endorphin in the anterior and intermediate lobes of the rat pituitary. J. supramolec. Struct. *8:* 247–262 (1978).
10 Eipper, B.A.; Mains, R.E., and Guenzi, D.: High molecular weight forms of adrenocorticotropic hormones are glycoproteins. J. biol. Chem. *251:* 4121–4126 (1976).
11 Gainer, H.; Sarne, Y., and Brownstein, M.J.: Biosynthesis and axonal transport of rat neurohypophysial proteins and peptides. J. Cell Biol. *73:* 366–381 (1977).
12 Gainer, H.; Sarne, Y., and Brownstein, M.J.: Neurophysin biosynthesis: conversion of a putative precursor during axonal transport. Science, N.Y. *195:* 1354–1356 (1977).
13 Habener, J.F. and Kronenberg, H.M.: Parathyroid hormone biosynthesis: structure and function of biosynthetic precursors. Fed. Proc. Fed. Am. Socs exp. Biol. *37:* 2561–2566 (1978).
14 Lingappa, V.R.; Katz, F.N.; Lodish, H.F., and Blobel, G.: A signal sequence for the insertion of a transmembrane glycoprotein. J. biol. Chem. *253:* 8867–8670 (1978).
15 Loh, Y.P.: Immunological evidence for two common precursors to corticotropins, endorphins, and melanotropin in the neurointermediate lobe of the toad pituitary. Proc. natn. Acad. Sci. USA *76:* 796–800 (1979).
16 Loh, Y.P. and Gainer, H.: The role of glycosylation in the biosynthesis, degradation and secretion of the ACTH-β-lipotropin common precursor and its peptide products. FEBS Lett. *96:* 269–272 (1978).
17 Mains, R.E.; Eipper, B.A., and Ling, N.: Common precursor to corticotropins and endorphins. Proc. natn. Acad. Sci. USA *74:* 3014–3018 (1977).
18 Mobley, W.C.; Seruer, A.C.; Ishi, D.N.; Riopelle, R.J., and Shooter, E.M.: Nerve growth factor. New Engl. J. Med. *297:* 1096–1104 (1977).
19 Nakanishi, S.; Inoue, A.; Kita, T.; Numa, S.; Chang, A.C.Y.; Cohen, S.N.; Nunberg, J., and Schimke, R.T.: Construction of bacterial plasmids that contain the nucleotide sequence for bovine corticotropin-β-lipotropin precursor. Proc. natn. Acad. Sci. USA *75:* 6021–6025 (1978).
20 Nakanishi, S.; Taii, S.; Hirata, Y.; Matsukura, S.; Imura, H., and Numa, S.: A large product of cell-free translation of messenger RNA coding for corticotropin. Proc. natn. Acad. Sci. USA *73:* 4319–4323 (1976).
21 Orenstein, N.S.; Dvorak, H.F.; Blanchard, M.H., and Young, M.: Nerve growth factor: a protease that can activate plasminogen. Proc. natn. Acad. Sci. USA *75:* 5497–5500 (1978).
22 Palade, G.: Intracellular aspects of the process of protein synthesis. Science, N.Y. *189:* 347–358 (1975).

23 Pickering, B.T. and Jones, C.W.: The neurophysins; in Li, Hormonal proteins and peptides, pp. 103–158 (Academic Press, New York 1978).
24 Roberts, J.L. and Herbert, E.: Characterization of a common precursor to corticotropin and β-lipotropin: cell-free synthesis of the precursor and identification of corticotropin peptides in the molecule. Proc. natn. Acad. Sci. USA *74:* 4826–4830 (1977).
25 Roberts, J.L. and Herbert, E.: Characterization of a common precursor to corticotropin and β-lipotropin: identification of β-lipotropin peptides and their arrangement relative to corticotropin in the precursor synthesized in a cell-free system. Proc. natn. Acad. Sci. USA *74:* 5300–5304 (1977).
26 Rubinstein, A.H.; Steiner, D.F.; Horowitz, D.L.; Mako, M.E.; Block, M.B.; Starr, J.I.; Kuzuya, H., and Melani, F.: Clinical significance of circulating proinsulin and C-peptide. Recent Prog. Horm. Res. *33:* 435–475 (1977).
27 Sachs, H.; Fawcett, Y.; Takabatake, Y., and Portanova, R.: Biosynthesis and release of vasopressin and neurophysin. Recent Prog. Horm. Res. *25:* 447–491 (1969).
28 Sachs, H. and Takabatake, Y.: Evidence for a precursor in vasopressin biosynthesis. Endocrinology *75:* 943–948 (1964).
29 Scott, A.P.; Ratcliffe, J.F.; Rees, L.H.; Landon, J.; Bennett, H.P.J.; Lowry, P.J., and McMartin, C.: Pituitary peptide. Nature new Biol. *244:* 65–67 (1973).
30 Silman, R.E.; Holland, D.; Chard, T.; Lowry, P.J.; Hope, J.; Robinson, J.S., and Thorburn, G.D.: The ACTH 'family tree' of the rhesus monkey changes with development. Nature, Lond. *276:* 526–528 (1978).
31 Steiner, D.F.: Insulin today. Diabetes *26:* 322–340 (1976).
32 Steiner, D.F.; Kemmler, W.; Tager, H.S., and Peterson, J.D.: Proteolytic processin in the biosynthesis of insulin and other proteins. Fed. Proc. Fed. Am. Socs exp. Biol. *33:* 2105–2115 (1974).
33 Steiner, D.F. and Oyer, P.E.: The biosynthesis of insulin and a probable precursor of insulin by a human islet cell adenoma. Proc. natn. Acad. Sci. USA *57:* 473–480 (1967).
34 Swanson, L.W.: Immunohistochemical evidence for a neurophysin-containing autonomic pathway arising in the paraventricular nucleus of the hypothalamus. Brain Res. *128:* 346–353 (1977).
35 Yalow, R.S. and Berson, S.A.: Characteristics of 'big ACTH' in human plasma and pituitary extracts. J. clin. Endocr. Metab. *36:* 415–523 (1973).

Dr. H. Gainer, Section on Functional Neurochemistry, Laboratory of Developmental Neurobiology, National Institute of Child Health and Human Development, National Institute of Health, Bethesda, MD 20205 (USA)

Prog. biochem. Pharmacol., vol. 16, pp. 69–83 (Karger, Basel 1980)

Studies of α-MSH-Containing Nerves in the Brain

T.L. O'Donohue and D.M. Jacobowitz

Laboratory of Clinical Science, National Institute of Mental Health, Bethesda, MD

Introduction

α-Melanocyte stimulating hormone (α-MSH) was recently shown to be present in neurons of the mammalian brain (5, 8, 15, 22). The presence of α-MSH-containing fibers in specific areas of the brain, in addition to known potent central nervous system effects (10, 13, 31), suggests a possible neuromodulator or neurotransmitter role for this peptide in brain function. The present communication will summarize the status of recent knowledge of intraneuronal α-MSH and provide further information concerning the anatomy and physiology of α-MSH in the brain.

Observations

Localization of α-MSH-Containing Nerves by Means of Immunofluorescence

An indirect immunohistochemical procedure (3, 7) was used with minor modifications as described previously (8). Rat brain sections were fixed with paraformaldehyde and incubated with α-MSH antiserum (1:75 dilution), and also α-MSH antiserum which had been incubated overnight with 0.1 ng/50 μl–1.0 μg/50 μl of α-MSH, ACTH and ACTH 1–10 (8). Immunoreactivity was prevented by 1 and also by 0.01 ng/ml of α-MSH. No decrease in immunoreactivity was noted with any of the other peptides used as preabsorbants.

α-MSH immunoreactivity was observed within discrete varicose nerve fibers and only faintly visible within neuronal perikarya in the arcuate nucleus. Treatment with vinblastine resulted in intensely fluorescent perikarya from the most anterior to the most caudal part of the arcuate nucleus. An extensive system of discrete varicose beaded nerve fibers, with positive α-MSH immunoreactivity, appeared to extend throughout the rat brain. A mapping of the distribution of

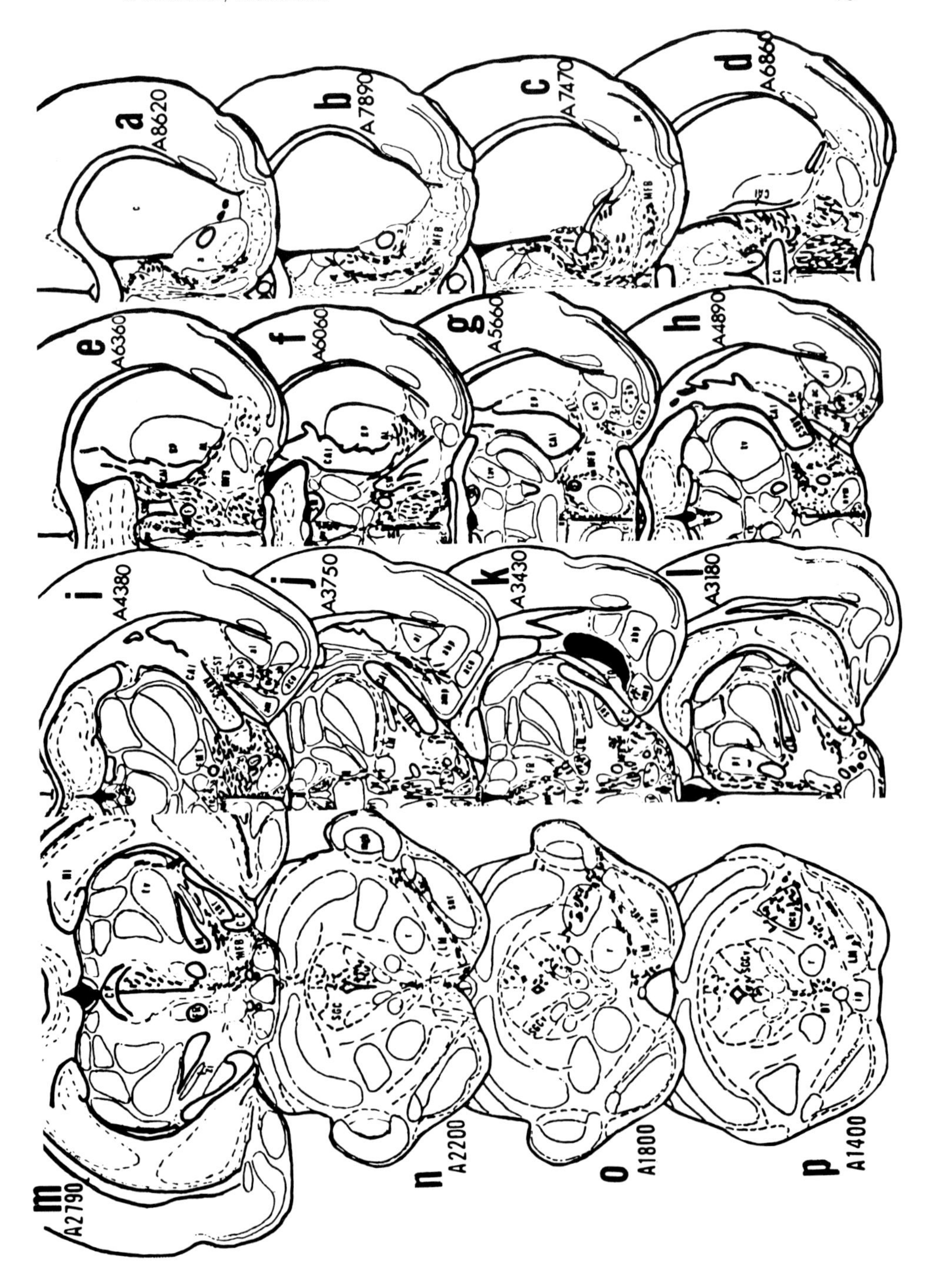
a
A8620
b
A7890
c
A7470
d
A6860
e
A6360
f
A6060
g
A5660
h
A4890
i
A4380
j
A3750
k
A3430
l
A3180
m
A2790
n
A2200
o
A1800
p
A1400

Fig. 2. A dense accumulation of varicose α-MSH fibers in the dorsomedial nucleus of the hypothalamus. The appearance of fine varicose nerves is typical of that seen throughout the rat brain (×370).

α-MSH immunoreactive fibers and perikarya is presented in figure 1. In the septal region, some α-MSH immunoreactive fibers coursed medial to the nucleus accumbens in the septohypothalamic tract and in the diagonal band, and appeared to terminate in the lateral septal nucleus and the dorsal and ventral parts of the interstitial nucleus of the stria terminalis. A dense population of α-MSH-containing fibers was observed in the medial preoptic nucleus and the periventricular (fig. 1d), dorsomedial (figs. 1i, 2), and anterior hypothalamic nuclei (figs. 1e, f). Some fibers were observed in the paraventricular (fig. 1g), lateral preoptic (fig. 1d) and posterior hypothalamic nuclei (figs. 1j–m). In the thalamus, a moderate fiber distribution was found in the rhomboid nucleus, and a heavy

---

Fig. 1. Schematic coronal sections of α-MSH-containing fibers in the rat brain. Coordinates are taken from the topographic atlases of *Jacobowitz and Palkovits* (9) and *Palkovits and Jacobowitz* (20). These, as well as references 8 and 15, should be consulted for abbreviations. α-MSH fibers are shown as dashed lines or dotted accumulations.

distribution of fibers in the thalamic periventricular nucleus (figs. 1e–l). Numerous fibers also appeared to project from the stria terminalis to innervate the amygdala (figs. 1e–k).

α-MSH immunoreactive fibers were also seen in the midbrain and hindbrain. Many fibers were observed in the mesencephalic central gray (figs. 1l–p) and a moderate number appeared to project between the central gray and the region of the cuneiform and dorsal parabrachial nuclei (figs. 1o–p). In the pons very few, if any, immunoreactive fibers were noted in the locus coeruleus, but some were observed in the dorsal lateral tegmental nucleus, dorsal parabrachial nucleus, nucleus of the mesencephalic tract, trigeminal nerve and the superior cerebellar peduncle. In the medulla, a moderate number of fibers were located in the nucleus of the solitary tract and a few fibers were seen in the reticular formation. No fibers were found in the spinal cord.

High Pressure Liquid Chromatography

The high pressure liquid chromatographic system (HPLC) consisted of a Waters Associates Model 204 Liquid Chromatograph equipped with a Model UK6 injector, two Model 6000A pumps, a Model 660 programmer, a Model 440 absorbance detector and a 4 mm × 30 cm column of μBondapak/C_{18}. A mobile phase of a mixture of de-gassed acetonitrile (Waters Associates) and ammonium acetate buffer (0.01 M, pH 4, filtered through a 0.45 Millipore filter and de-gassed) was chosen because of previous demonstration of its adequacy in resolving peptides (24). Gradient chromatography was initially performed to determine appropriate solvent ratios for α-MSH elution. A solvent system of 28% acetonitrile in ammonium acetate (28% CH_3CH/NH_4COOCH_3) was chosen and run isocratically at a rate of 2 ml/min.

Whole brains, hypothalami, and standard α-MSH in 1% egg albumen in phosphate-buffered saline at pH 7.4 (PBSEA) were each boiled in 2N acetic acid for 10 min and homogenized by sonication. Samples were centrifuged at 8,000 × g for 5 min and aliquots of the supernatant were removed and lyophilized. Lyophilates were resuspended in 28% CH_3CN/NH_4COOCH_3 and centrifuged at 8,000 × g for 5 min to remove particulate matter. Aliquots of the supernatant were then injected onto the column. Fractions were collected every 15 sec between 1 and 6 min following injection, and thereafter every 1 min between 6 and 10 min. Aliquots of these samples were concentrated in a vacuum centrifuge and assayed to determine the quantity of immunoreactive α-MSH.

As determined by absorbance at 280 nm, α-MSH elutes from the column and through the detector with a peak at 3.6 min after injection onto the high pressure liquid chromatographic (HPLC) column (fig. 3). Fractions collected from the HPLC show a peak range of standard α-MSH immunoreactivity eluting from the HPLC from 3.5 to 4.25 min after injection, with a peak at approximately 3.75 min (fig. 3). As shown in figure 3, whole brain and hypothalamic extracts each have a peak of immunoreactivity between 3.5 and 4.5 min, exactly the same as the retention time for standard α-MSH, as well as a peak of immunoreactivity between 2.5 and 3.0 min. In whole brain a smaller immunoreactive peak also exists between 4.75 and 5.25 min.

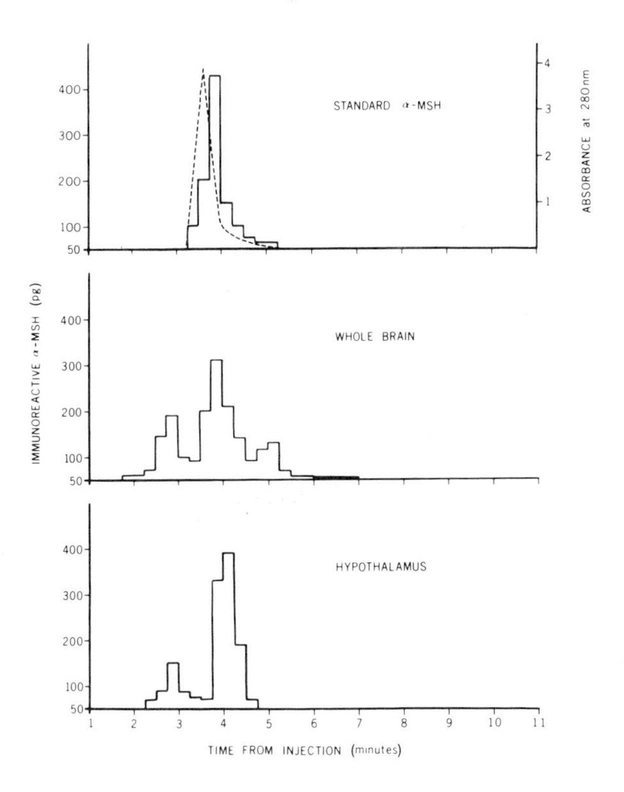

Fig. 3. High pressure liquid chromatographs of standard α-MSH and hypothalamic and whole brain extracts. Samples eluting from the column were collected every 15 sec between 1 and 6 min after injection. The solvent was evaporated and samples were assayed for α-MSH by radioimmunoassay. The retention time of standard α-MSH was determined by absorbance at 280 nm (0.01 absorbance units full scale on the right ordinate) and by immunoreactivity (left ordinate). The major α-MSH immunoreactive peak of whole brain and hypothalamic extracts was between 3.5 and 4.5 min, exactly the same as standard α-MSH. Two additional peaks were observed in whole brain, with retention times of 2.75 and 5.25 min. One additional peak, with a retention time of 2.75 min, was observed in hypothalamic extracts.

Radioimmunoassay of α-MSH Concentration in Discrete Brain Regions

Microdissection of the brain. Rats were killed by decapitation and the brains were rapidly removed, mounted on specimen plates and frozen on dry ice. Alternate 300 μm and 60 μm coronal sections for microdissection and histology respectively were cut on a cryostat at −8°C. The 300 μm sections were then microdissected by the technique of *Palkovits* (19). The approximate sizes and coordinates of the dissected regions have been previously described (15).

Table I. Microdissection parameters and α-MSH concentrations (pg/μg protein ± S.E.M.) in discrete regions of rat and one human brain

Region	Normal	Rat		Human
		arcuate nucleus lesion		
		sham	lesion	
Piriform cortex	0.20 ± 0.09			
Hippocampus	0.25 ± 0.05			0
Caudate nucleus (n.)	0.13 ± 0.03			
Globus pallidus	0.39 ± 0.15			
N. accumbens	0.33 ± 0.05			7.57
Septum dorsalis	0.15 ± 0.07			
Septum lateralis	1.29 ± 0.16	1.2 ± 0.30	0.16 ± 0.16[1]	
N. stria terminalis, dorsal	1.43 ± 0.27	1.8 ± 0.25	0.15 ± 0.15[1]	4.77
N. stria terminalis, ventral	4.52 ± 0.41			
Olfactory tubercle	0.20 ± 0.03			2.00
N. tractus diagonalis	1.06 ± 0.32	0.76 ± 0.13	0.09 ± 0.09[1]	3.00
Medial preoptic n.	3.75 ± 0.32	4.17 ± 0.50	n.d.[1]	5.51
Lateral preoptic n.	0.97 ± 0.06			
Suprachiasmatic n.	1.01 ± 0.23			
Anterior hypothalamic n.	3.49 ± 0.55	4.65 ± 0.75	0.27 ± 0.27[1]	3.75
Periventricular n.	7.19 ± 0.73	7.23 ± 0.49	0.25 ± 0.25[1]	
Supraoptic n.	1.95 ± 0.41			9.34
Paraventricular n.	5.75 ± 0.35	4.93 ± 0.35	n.d.[1]	
Arcuate n.	8.72 ± 1.33			
Median eminence	11.02 ± 1.60			
Ventromedial n.	3.23 ± 0.63			5.71
Dorsomedial n.	8.95 ± 0.65			5.95
Posterior hypothalamic n.	4.32 ± 0.56			
Ansa lenticularis	3.83 ± 0.56			

n for rat = 5–7; [1] Statistically significant ($p < 0.01$).

Extraction. Micropunches were delivered into 100 μl of 2 N acetic acid in plastic microtubes (0.4 ml) at 4°C. Samples were subsequently boiled for 10 min and then homogenized by sonication. A 10–20 μl aliquot was removed for protein determination (12). Samples were then centrifuged at 8,000 × g for 5 min and aliquots of the supernatant were removed and lyophilized in 12 × 75 mm borosilicate glass test tubes. Preliminary experiments demonstrated that this technique extracted essentially all of a standard amount of α-MSH added to brain tissue in the acid supernatant. Negligible amounts of α-MSH remained in the 8,000 × g pellet.

α-MSH radioimmunoassay. α-MSH in brain was determined by a modification of several procedures (18, 29, 30). Samples were always assayed in duplicate, and initially three sets of duplicates with separate aliquots were assayed to demonstrate parallel displacement of α-MSH when compared to a standard curve. Synthetic α-MSH was labeled with ^{125}I using

Table I. (continued)

Region	Normal	Rat		Human
		arcuate nucleus lesion		
		sham	lesion	
Medial forebrain bundle				
rostral	2.34 ± 0.37			
caudal	3.99 ± 0.72			
Anterior ventral thalamic n.	0.42 ± 0.18			
Ventral thalamic n.	0.65 ± 0.11			
Rhomboid n.	4.56 ± 0.60			
Periventricular n. thalamus	7.19 ± 0.73	6.98 ± 0.75	0.17 ± 0.17[1]	
Parafascicular n.	1.34 ± 0.25			
Medial amygdaloid n.	0.90 ± 0.13			
Cortical amygdaloid n.	1.13 ± 0.30			
Central amygdaloid n.	1.93 ± 0.19			
Basal amygdaloid n.	1.73 ± 0.16			
Central gray, rostral	0.48 ± 0.09			
Central gray, caudal	3.44 ± 0.67			
Supramammillary decussation	0.36 ± 0.07			
Superior colliculus	0.23 ± 0.10			
Inferior colliculus	0.31 ± 0.09			
Interpeduncular n.	0.19 ± 0.10			
Cuneiform n.	1.27 ± 0.26			
Dorsal raphe	4.15 ± 0.87			
Medial raphe	1.96 ± 0.33			
Lateral dorsal tegmental n.	1.49 ± 0.27			
N. tractus solitarius	1.07 ± 0.27			

n for rat = 5–7; [1] Statistically significant ($p < 0.01$).

the chloramine-T method of *Greenwood et al.* (6) and purified using QAE Sephadex column chromatography. The antibody to α-MSH was generously provided by Drs. *M.C. Tonon* and *H. Vaudry,* Laboratoire d'Endocrinologie, Mont-Saint-Aignan, France. *Vaudry et al.* (30) have demonstrated the high specificity of this antibody to α-MSH. All steps of the assay were performed at 4°C. On the first day, lyophilates, standards and blanks were resuspended in a volume of 400 μl of PBSEA in 12 × 75 mm borosilicate glass test tubes, and 50 μl of diluted α-MSH antibody was added to achieve a final assay dilution of 1:40,000. On day 2, approximately 3,000 cpm (4–8 pg) of the ^{125}I-labeled α-MSH in 50 μl of 25 mM Tris buffer containing 0.2% albumin was added to each tube. On day 4, 200 μl of 1% normal rabbit serum in PBS, pH 7.6, and 200 μl of PBS containing 50 μl of sheep antiserum against rabbit γ-globulin were added simultaneously. On the sixth day, samples were centrifuged at 5,000 × g for 30 min, supernatants aspirated, and the radioactivity in the pellet measured.

Extracts of microdissected regions of the rat brain were immunologically similar to synthetic α-MSH as demonstrated by parallel displacement of the ^{125}I-labeled α-MSH. α-MSH was unevenly distributed in individual nuclei and regions throughout the rat brain, as shown in table I. The highest concentrations of α-MSH were observed in the hypothalamic region, and were 10-fold greater than the lowest concentrations. A 70- to 100-fold range exists when comparing the lowest values of α-MSH detected in the brain with the highest values in the median eminence and the dorsomedial, arcuate, and periventricular nuclei.

Extrahypothalamic regions also contained varying amounts of α-MSH. Moderate α-MSH concentrations were detected in the nuclei of the septum, amygdala, thalamus, midbrain and hindbrain. A particularly high concentration was observed in the periventricular nucleus of the thalamus. Low or non-detectable concentrations of α-MSH were found in the cortex, striatum, nucleus accumbens, olfactory tubercle, hippocampus, medial geniculate body, substantia nigra reticulata, colliculi, interpeduncular nucleus and the locus coeruleus.

Microdissection of discrete nuclei of a single human brain (26-year-old female, cystic fibrosis, 14 hours post-mortem) revealed that even greater quantities of α-MSH were contained in numerous nuclei. No α-MSH was detected in the cortex or hippocampus (table I).

Effect of Arcuate Lesions on the Regional Distribution of α-MSH

Lesions of the arcuate nucleus were performed (see 15 for details) in order to determine the contribution of this nucleus to other α-MSH-containing areas, since this was the only region that contained α-MSH-positive cell bodies. Rats were killed 6 days after surgery. The damage to the arcuate nucleus and surrounding regions was determined by histological examination. Most or all of the arcuate nucleus was destroyed in 5 of the 6 lesioned rats. In these animals, the α-MSH concentrations in all brain regions examined were significantly depleted compared to control animals, and were not different from values of 0.0 pg of α-MSH as determined by analysis of variance (table I).

Circadian Rhythm of α-MSH in Discrete Brain Areas

Many functions of various organs (e.g., pineal) undergo diurnal rhythmic variations. Little information is available on possible intrinsic oscillations of α-MSH in discrete brain regions, although a diurnal rhythm has been reported in the rat pituitary (27) and blood (29). It was therefore of interest to study α-MSH concentrations in discrete areas of the brain over a 24-h period.

Rats were killed by decapitation at 4 hourly intervals over a 20-h period, starting at 0900 hours. The brains were removed and processed for microdissection and radioimmunoassay as described above.

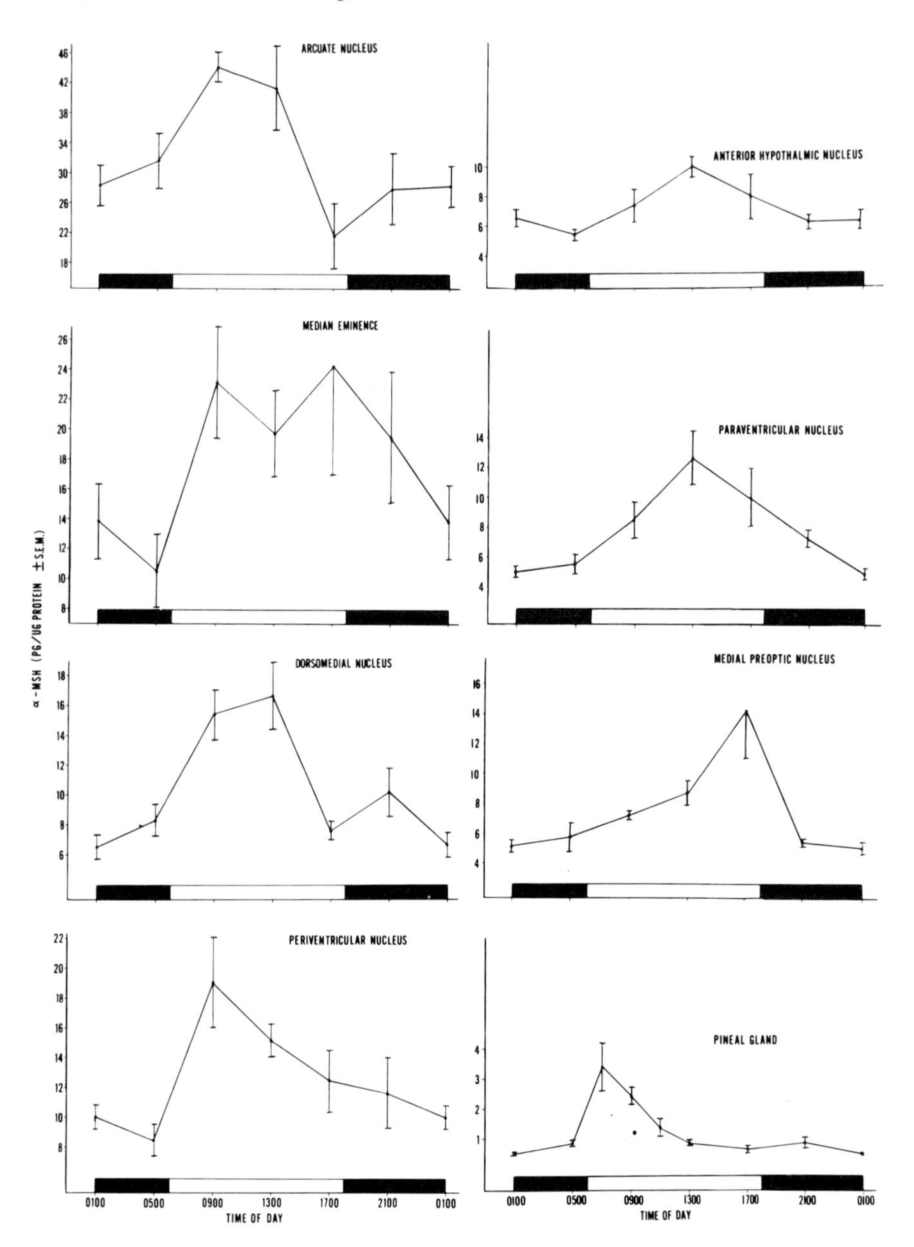

Fig. 4. α-MSH concentrations throughout a 24-h light-dark cycle in discrete regions of the rat brain and pineal gland. N = 6–8 per region. A significant ($p < 0.05$) diurnal rhythm of α-MSH was observed in all regions investigated.

A significant diurnal variation of α-MSH was observed in all brain regions studied (fig. 4). In each case, the highest α-MSH concentrations occurred during the day and the lowest at night. The time of the actual peak of α-MSH concentration did, however, vary from region to region during the light part of the cycle. The peak concentrations of α-MSH in the arcuate nucleus, median eminence, dorsomedial nucleus and periventricular nucleus of the thalamus were all seen at 0900 hours. In the arcuate nucleus, the peak α-MSH concentration at 0900 was significantly higher than the concentrations at 0100, 0500, 1700 and 2100 hours. The arcuate nucleus α-MSH concentration at 1300 was also significantly higher than at 0100, 1700 and 2100 hours. Peak concentrations in the median eminence, between 0900 and 1700, were twice as high as the nadir values at 0500. The highest concentrations of α-MSH in the dorsomedial nucleus were at 0900 and 1300 and were significantly higher than concentrations at 0100, 0500, 1700 and 2100 hours. Similarly, the peak α-MSH concentration in the periventricular nucleus of the thalamus was at 0900 and was significantly higher than the levels at 0100, 0500, 1700 and 2100 hours.

The peak α-MSH concentrations in the anterior hypothalamic and paraventricular nuclei both occurred at 1300 hours, slightly later than seen in the previous nuclei. The peak concentration at 1300 hours in the anterior hypothalamic nucleus was significantly higher than the levels at 0100, 0500, and 2100 hours. Similarly, at 1300 hours, paraventricular concentrations were significantly higher than at 0100, 0500, 0900, and 2100 hours. The peak α-MSH concentration in the medial preoptic nucleus occurred at 1700 hours, later than the peak observed in other nuclei. The α-MSH concentrations at this time were significantly higher than those at 0100, 0500, 0900, 1300, and 2100 hours.

Circadian Rhythm of α-MSH in the Pineal Gland

Previously described antagonistic interactions between α-MSH and melatonin in mediating physiological and behavioral circadian rhythms (11) suggest the possibility of an α-MSH involvement in the circadian processes of the pineal gland. To investigate this possibility pineal α-MSH was identified and characterized by HPLC combined with radioimmunoassay, and the influences of different lighting regimens, hypophysectomy and superior cervical ganglionectomy on immunoreactive α-MSH were investigated.

HPLC techniques demonstrated that the major α-MSH-immunoreactive compound has the same retention time as standard α-MSH. A minor peak of immunoreactivity eluted just before that of standard α-MSH.

A circadian variation of pineal α-MSH concentrations was observed during the months of July and August (fig. 4). Concentrations of α-MSH peaked at 0700, one hour after the lights were turned on, and returned to baseline concentrations by 1300 hours. The concentrations at 0700 and 0900 were significantly higher than concentrations at all other times of the day.

Animals kept in darkness for one week demonstrated a significant rhythm in pineal α-MSH concentrations. The peak α-MSH concentration in these animals was at 0500, whereas in rats kept in normal light-dark conditions it occurred at 0700. This peak was also significantly higher than the peak observed at 0700 in animals kept in normal light-dark. The group of rats which were moved to a lighted room at 0600 on the seventh day of chronic darkness showed an α-MSH variation identical with that of the chronic dark rats. In contrast to the chronic dark group, the pineal α-MSH rhythm in animals housed in chronic light was significantly suppressed.

Effect of Hypophysectomy on Pineal α-MSH

Concentrations of pineal α-MSH are similar at 0500 and 0900 hours in normal rats and rats hypophysectomized for 2 weeks. There is also a significant increase in α-MSH concentrations at 0900 compared to that at 0500 in both groups of animals.

Effect of Superior Cervical Ganglionectomy on Pineal α-MSH

Bilateral superior cervical ganglionectomized and sham operated rats had statistically identical α-MSH rhythms. It is interesting to note, however, that the pineal α-MSH concentrations in both these groups were higher, and seemed to increase earlier (0500 vs 0700), than those observed in the previous studies. One major difference in these two groups is that the first set of experiments was performed during the months of July and August and the ganglionectomy study took place in December.

Discussion

The use of immunohistochemistry in conjunction with immunochemical assays, surgical and specific lesion methods, provides an otherwise unattainable level of resolution for mapping peptidergic neurons. The present studies demonstrate that the data obtained by radioimmunoassay and by immunohistochemistry are complementary. In general, good agreement exists between the immunohistochemical results and the immunoassay of α-MSH in discrete areas of the brain. The immunofluorescence study has provided information concerning the discrete localization of immunoreactive material, while the radioimmunoassay provided for the provisional identification of the immunoreactive material. Greater confidence for the identification of α-MSH in brain tissue was gained by high pressure liquid chromatography. With immunoassays, parallelism of dose-response curves was ascertained by using extracts of several hypothalamic nuclei. With HPLC, the major peak of immunoreactivity eluted in the same position as a sample of synthetic α-MSH that was applied subsequently to the column. Lesions of the arcuate nucleus demonstrated that this cell body grouping was solely responsible for the α-MSH projection pathways. This, plus the fact that the hypophysectomized rats showed no detectable decrease in the number of α-MSH nerves in the brain (5, 8, 15; but also see 14, 18), demonstrates that the pituitary gland is an unlikely source of α-MSH in the brain.

The immunocytochemical and radioimmunochemical data reveal a regional distribution of immunoreactive α-MSH throughout the rat and human brain. The highest concentrations of α-MSH and densities of the α-MSH-containing fibers were found in the preoptic area and hypothalamus as well as in particular nuclei of the septum and thalamus. No α-MSH nerves were observed in the cerebral cortex, cerebellum, hippocampus or spinal cord.

The general circadian pattern of high α-MSH levels during the day and low concentrations at night was observed in all brain regions examined. A similar circadian variation was described in the rat pituitary (27) and blood (29). In the pituitary-portal system, therefore, increased pituitary concentrations are related to increased α-MSH secretion. Whether an increased concentration of α-MSH in the brain represents an increased or decreased α-MSH release can not be determined from the present data.

It is of interest that the daytime peak concentration of α-MSH in the pineal is correlated with the evening peak height of melatonin in this organ (see 23). Furthermore, the amplitudes of both circadian rhythms are increased in chronic dark and decreased in chronic light. However, in contrast to the melatonin effect, superior cervical ganglionectomy (see 23, 28) does not alter the α-MSH cyclicity, which suggests that norepinephrine does not influence α-MSH rhythmic activity in the pineal gland. Since it has been suggested that α-MSH inhibits melatonin biosynthesis (11), it is tempting to speculate that the α-MSH may modulate the well-known pineal and melatonin inhibitory effects of gonadal function.

The role of the circadian variation of α-MSH in the mammal is unknown but administration of α-MSH to rats and man produces potentially adaptive behavioral effects, such as increased attention, arousal and vigilance (4, 10, 13, 21, 31). An α-MSH effect of sleep-walking activity has also been reported in rats (21). Both sleep-walking behavior and attention processes demonstrate an obvious circadian rhythmic pattern. Furthermore, *Sandman et al.* (25) have demonstrated that α-MSH improves a passive avoidance response in the dark but not in the light period of the light cycle. These data may indicate a role for α-MSH in circadian variations in behavioral processes. Due to the fact that α-MSH in rat brain is contained in many regions which have been implicated in the neuronal processing of arousal or motivated behaviors, the variations of α-MSH concentrations in these brain areas may indicate a neuromodulator or neurotransmitter role for α-MSH in circadian mediation of behavior.

Summary

α-Melanocyte stimulating hormone (α-MSH) immunofluorescence was observed in the rat brain using a highly specific and well-characterized antibody. α-MSH was contained in the arcuate nucleus cell bodies and in varicose fibers distributed throughout the brain stem. α-MSH-containing fibers were present in various nuclei of the hypothalamus, preoptic area, septum, amygdala, mammillary body and central gray area. The distribution of α-MSH was verified by radioimmunoassay following microdissection of discrete brain nuclei. High concentrations of α-MSH were contained in the median eminence, medial preoptic, anterior hypothalamic, periventricular, paraventricular, arcuate, dorsomedial, posterior hypothalamic nuclei and bed nucleus of the stria terminalis. Moderate α-MSH concentrations were noted in the amygdala, septal area, central gray, dorsal raphe, and the nucleus tractus solitarius.

Hypophysectomy did not significantly reduce the quantity of α-MSH fibers in the brain, thereby suggesting an extra-pituitary source of α-MSH. Lesions of the arcuate nucleus did, however, completely abolish α-MSH-like immunoreactivity.

The α-MSH-like compound in the brain has immunochemical and electrophoretic properties similar to those of standard α-MSH. High pressure liquid chromatographic analysis demonstrated that the α-MSH immunoreactivity in the brain was comprised of one major component having a retention time identical with that of standard α-MSH, as well as 2 minor components.

In male rats kept on a 12-h light-dark schedule (0600–1800 hours), there was a diurnal rhythm of α-MSH in the hypothalamic nuclei with peak content at 0900 in the arcuate and periventricular nuclei of the thalamus; at 1300 in the dorsomedial, paraventricular and anterior hypothalamic nuclei, and at 1700 in the medial preoptic nucleus.

In the pineal gland a diurnal rhythm was also observed with a peak concentration at 1900. Six days of constant light, however, abolished the morning rise in α-MSH. Rats kept in constant dark for 6 days showed a marked increase in the α-MSH peak (12-fold) occurring at 0500. Normal diurnal rhythm of α-MSH was still observed in hypophysectomized rats.

It is suggested that α-MSH may function as a neurotransmitter or as a neuromodulator in the brain. The extensive distribution of α-MSH in the brain suggests that it is involved in significant neuronal circuitry and supports the notion of a neuroregulatory role for this neuropeptide. This lays the groundwork for a rational approach to further study of possible interactions between α-MSH and other neuronal systems.

References

1 Arbit, J.: Diurnal cycles and learning in earthworms. Science, N.Y. *126:* 654–655 (1957).

2 Battig, K.: Drug effects on exploration on a combined maze and open field system by rats. Ann. N.Y. Acad. Sci. *159:* 880–897 (1969).

3 Coons, A.H.: Fluorescent antibody methods; in Danielli, General cytochemical methods, pp. 399–422 (Academic Press, New York 1958).

4 de Weid, D.: Pituitary control of avoidance behavior; in Martini, Motta and Fraschini, The hypothalamus, pp. 1–8 (Academic Press, New York 1970).

5 Dube, D.; Lissitzky, J.C.; Leclerc, R., and Pelletier, G.: Localization of α-melanocyte stimulating hormone in rat brain and pituitary. Endocrinology *102:* 1283–1291 (1978).

6 Greenwood, F.C.; Hunter, M.M., and Glover, J.S.: The preparation of ^{131}I-labeled human growth hormone of high specific radioactivity. Biochem. J. *89:* 114–123 (1963).
7 Hökfelt, T.; Fuxe, K., and Goldstein, M.: Applications of immunohistochemistry to studies on monoamine cell systems with special reference to nervous tissues. Ann. N.Y. Acad. Sci. *254:* 407–432 (1975).
8 Jacobowitz, D.M. and O'Donohue, T.L.: α-Melanocyte stimulating hormone: Immunocytochemical identification and mapping in neurons of the rat brain. Proc. natn. Acad. Sci. USA *75:* 6300–6304 (1978).
9 Jacobowitz, D.M. and Palkovits, M.: Topographic atlas of catecholamine and acetylcholinesterase-containing neurons in the brain. I. Forebrain (telencephalon, diencephalon). J. comp. Neurol. *157:* 13–28 (1974).
10 Kastin, A.J.; Plotnikoff, N.P.; Schally, A.V., and Sandman, C.A.: Endocrine and CNS effects of hypothalamic peptides and MSH; in Ehrenpreis and Kopin, Reviews of neuroscience, pp. 111–148 (Raven Press, New York 1976).
11 Kastin, A.J.; Viosca, S.; Nair, R.M.G.; Schally, A.V., and Miller, M.C.: Interactions between pineal, hypothalamus and pituitary involving melatonin, MSH release-inhibiting factor and MSH. Endocrinology *91:* 1323–1328 (1973).
12 Lowry, O.; Rosebrough, M.; Farr, A., and Randall, R.: Protein measurement with the Folin phenol reagent. J. biol. Chem. *193:* 265–275 (1951).
13 Miller, L.H.; Kastin, A.J., and Sandman, C.A.: Psychobiological actions of MSH in man; in Tilders, Swaab and van Wimersma Greidanus, Melanocyte stimulating hormone: control, chemistry and effects, pp. 153–161 (Karger, Basel 1977).
14 O'Donohue, T.L.; Holmquist, G.E., and Jacobowitz, D.M.: Effect of hypophysectomy on α-melanocyte stimulating hormone in discrete regions of the rat brain. Neuroscience Lett. (in press).
15 O'Donohue, T.L.; Miller, R.L., and Jacobowitz, D.M.: Identification, characterization and stereotaxic mapping of intraneuronal α-melanocyte stimulating hormone-like immunoreactive peptides in discrete regions of the rat brain. Brain Res. (in press).
16 O'Donohue, T.L.; Miller, R.L.; Pendleton, R.C., and Jacobowitz, D.M.: A diurnal rhythm of immunoreactive α-melanocyte stimulating hormone in discrete regions of the rat brain. Neuroendocrinology (in press).
17 O'Donohue, T.L.; Miller, R.L.; Pendleton, R.C., and Jacobowitz, D.M.: Demonstration of an endogenous circadian rhythm of α-melanocyte stimulating hormone in the rat pineal gland. Brain Res. (in press).
18 Oliver, C. and Porter, J.C.: Distribution and characterization of α-melanocyte-stimulating hormone in the rat brain. Endocrinology *102:* 697–705 (1978).
19 Palkovits, M.: Isolated removal of hypothalamic nuclei for neuroendocrinological and neurochemical studies; in Stumpf and Grant, Anatomical neuroendocrinology, pp. 72–80 (Karger, Basel 1975).
20 Palkovits, M. and Jacobowitz, D.M.: Topographic atlas of catecholamine and acetylcholinesterase-containing neurons in the brain. II. Hindbrain (mesencephalon, rhombencephalon). J. comp. Neurol. *157:* 29–42 (1974).
21 Panksepp, J.; Reilly, P.; Bishop, P.; Meeker, R.B., and Vilberg, T.R.: Effects of α-MSH on motivation, vigilance, and brain respiration. Pharmac. Biochem. Behav. *5,* Suppl. *1:* 59–64 (1976).
22 Pelletier, G. and Dube, D.: Electron microscopic immunohistochemical localization of α-MSH in the rat brain. Am. J. Anat. *150:* 201–205 (1977).
23 Quay, W.B.: Pineal chemistry in cellular and physiological mechanisms (Thomas, Springfield 1974).

24 Rivier, J.; Lazarus, L.; Perrin, M., and Brown, M.: Neurotensin analogues. Structure-activity relationships. J. med. Chem. *20:* 1409–1412 (1977).
25 Sandman, C.A.; Kastin, A.J., and Schally, A.V.: Behavioral inhibition as modified by melanocyte-stimulating hormone (MSH) and light-dark conditions. Physiol. Behav. *6:* 45–48 (1971).
26 Stephens, G.; McGaugh, J.L., and Alpern, H.P.: Periodicity and memory in mice. Psychonom. Sci. *8:* 201–202 (1967).
27 Tilders, F.J.H. and Smelik, P.G.: A diurnal rhythm in melanocyte-stimulating hormone content of the rat pituitary gland and its independence from the pineal gland. Neuroendocrinology *17:* 296–308 (1975).
28 Tomatis, M.E. and Orias, R.: Changes in melatonin concentration in pineal gland of rats exposed to continuous light or darkness. Acta physiol. Lat. Am. *17:* 227–233 (1967).
29 Usategui, R.; Oliver, C.; Vaudry, H.; Lombardi, G.; Rozenberg, I., and Mourre, A.M.: Immunoreactive α-MSH and ACTH levels in rat plasma and pituitary. Endocrinology *98:* 189–196 (1976).
30 Vaudry, H.; Tonon, M.C.; Delarue, R.; Vailant, R., and Kraicer, J.: Biological and radioimmunological evidence for melanocyte stimulating hormone (MSF) of extrapituitary origin in the rat brain. Neuroendocrinology *27:* 9–24 (1978).
31 Wimersma Greidanus, T.B. van: Effects of MSH and related peptides on avoidance behavior in rats; in Tilders, Swaab and van Wimersma Greidanus, Melanocyte stimulating hormone: Control chemistry and effects, pp. 129–139 (Karger, Basel 1977).

Dr. D.M. Jacobowitz, Laboratory of Clinical Science,
National Institute of Mental Health, Bethesda, MD 20205 (USA)

Prog. biochem. Pharmacol., vol. 16, pp. 84–94 (Karger, Basel 1980)

Substance P and Analogues: Biological Activity and Degradation

Vivian I. Teichberg[1] *and Shmaryahu Blumberg*[2]

Departments of Neurobiology[1] and Biophysics,[2] The Weizmann Institute of Science, Rehovot

Stimulated by the elucidation in 1971 of the chemical structure of Substance P (SP) (6a, 6b) an ever increasing number of studies have since been devoted to the understanding of the physiological role of this undecapeptide and to its assignment as a neurotransmitter.

In the peripheral nervous system, the hypotensive, vasodilatatory and smooth muscle contracting properties of SP, already acknowledged 49 years ago by *von Euler and Gaddum* (7), have been confirmed using synthetic SP but so far no evidence has been produced to correlate these properties of SP with a role in neurotransmission.

SP is found in nerve bundles along blood vessels, the gastrointestinal tract and various smooth muscles (10) but the existence of synapses where SP would be stored and released, elicit a postsynaptic action and be degraded, remains to be demonstrated.

In the central nervous system, SP seems to meet a number of criteria needed to establish its involvement in neurotransmission. For instance, SP has been localized in cell bodies, axons and nerve terminals (19) in the dorsal horn of the spinal cord and has been found to be released upon stimulation of the dorsal roots (17). In this area, SP was also reported to excite or facilitate the response of neurons to noxious stimulation (9). These findings are indeed compatible with the idea proposed by *Lembeck* in 1953 (15) that SP might be a sensory neurotransmitter. However, a definite conclusion has still to await the solution of the two following questions: 1) Is the postsynaptic action of SP identical to that of the natural neurotransmitter? 2) Is there a specific mechanism of inactivation of the SP response at a SP synapse?

The first question could be answered by the use of specific SP antagonists. So far, none of the proposed compounds (18) has been found to show a strict specificity to the SP receptor, and as a result the search for SP antagonists remains a most important challenge. The second question has been approached by several authors (1–3, 12) who reported on the degradation of SP by various

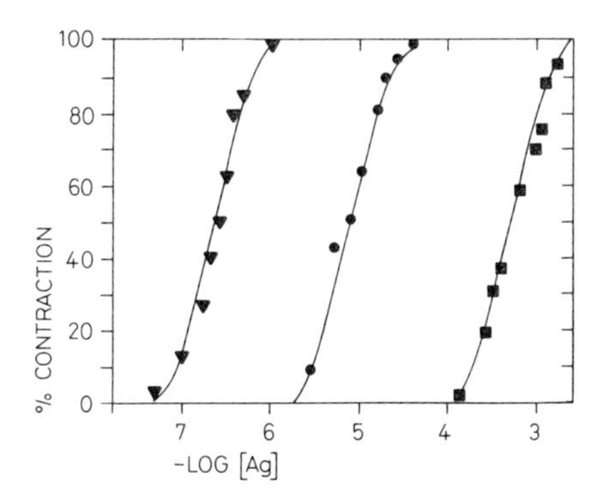

Fig. 1. Dose-response curves for the (9–11) tripeptide (■), (8–11) tetrapeptide (●) and (7–11) pentapeptide (▼). [Ag] stands for agonist concentration.

tissue proteases. However, the properties of these proteases do not appear compatible with those expected to exist at a SP synapse.

Addressing ourselves to these two questions we have undertaken a study of SP and SP analogues in order to characterize both the SP receptor and the enzymes involved in the specific degradation of SP. Our basic assumption is that the relationship between structure-function-stability of SP and its synthetic analogues should provide not only an understanding of the interaction of SP with its receptor and with its degrading enzyme(s) but should greatly facilitate the design of probes of the SP receptor including that of SP antagonists.

As a starting point in structure-function studies, we first investigated the possibility that not all the eleven amino acids composing SP were needed for an optimal interaction with the SP receptor. We therefore synthesized by stepwise synthesis, starting from the C-terminal methionine amide residue, a series of partial sequences and fragments of SP including the C-terminal heptapeptide, octapeptide and SP itself. We then tested the biological activity of these peptides by taking advantage of their ability to contract the guinea pig ileum.

The SP analogues synthesized were all capable of contracting the ileum in a reversible and dose-dependent fashion. For each peptide, an ED_{50} value can be defined and corresponds to the concentration of peptide causing a 50% contraction. At their highest concentrations, all peptides caused the same maximal contraction and their dose-response curves were parallel (fig. 1).

In table I, we summarize the $p(ED_{50})$ values obtained for various partial sequences of SP and for SP analogues. The smallest peptide fragment displaying contracting activity was Gly-Leu-$MetNH_2$. Leu-$MetNH_2$ and $MetNH_2$ did not show any activity at concentrations of 20 mM and 40 mM respectively and were not found to possess antagonistic properties. The increase in chain length has a marked effect on the contracting activity up to a length of 6–7 amino acid residues. As seen in table I, the increase in chain length beyond the C-terminal heptapeptide has little or no effect. Blocking the N-terminal residue with a

Table I. Contracting activity of Substance P and SP analogues measured on the guinea pig ileum assay

Peptide	$p(ED_{50}) \pm 0.2$	
	amino terminal free peptide	amino terminal blocked peptide with t-Boc
Substance P	8.6	N.D.
(4–11) Octapeptide	8.6	8.6
(5–11) Heptapeptide	8.6	8.6
(6–11) Hexapeptide	8.3	8.5
(7–11) Pentapeptide	6.6	7.7
(8–11) Tetrapeptide	5.0	N.D.
(9–11) Tripeptide	3.3	N.D.
(10–11) Dipeptide	no activity	

N.D. – Not determined.

tert-butyloxycarbonyl group significantly improves the potency of the pentapeptide but this effect is not marked for longer peptides. It can therefore be concluded that the SP receptor accommodates probably not more than the C-terminal hexa- or heptapeptide portion of SP. This conclusion is in total agreement with that reached by previous studies (5, 14, 20, 22). The question then arises whether the amino acid sequence of the hexa-heptapeptide is optimal in triggering the biological response. Can one improve its activity?

To answer this question, we studied in particular the effects resulting from the substitution of the glutamine residue in position 6 (from the C-terminal end) for other groups displaying various space-filling properties. This position was chosen because of the marked increase in contracting activity observed upon elongation of the C-terminal pentapeptide by one amino acid residue. The results of such substitutions are included in table II. The most effective substitute for glutamine is the pyroglutamyl residue whereas the least effective is the D-Ala residue.

Table II also lists the potency of peptides in which substitutions at other positions were carried out. Inspection of this table led us to the conclusion that the optimal size and sequence of SP analogues indeed correspond to the natural sequence of the C-terminal hexa-heptapeptide.

Having established the structure-function relationship between the various partial sequences of SP, we undertook a study of their stability. It has been known for many years, since the early work of *Gullbring* in 1943 (8), that Substance P is rapidly degraded by tissue extracts. (This might explain the fact that no reliable binding studies with radiolabeled SP have been carried out so far

Table II. Contracting activity of SP analogues and modified sequences measured on the guinea pig ileum

Peptide	$p(ED_{50}) \pm 0.2$
Boc-D-Ala Phe Phe Gly Leu $MetNH_2$	8.0
pGlu Phe Phe Gly Leu $MetNH_2$	8.6
D-Ala Phe Phe Gly Leu $MetNH_2$	7.3
Boc-Phe Phe Gly Leu $MetNH_2$	7.7
Boc-Gln Phe Phe D-Ala Leu $MetNH_2$	7.2
Boc-Gln Phe Phe β-Ala Leu $MetNH_2$	7.0

– see however 16.) Several authors (1–3, 12) have reported on the hydrolysis of SP by soluble brain enzymes but their observations do not seem to be directly related to the process of inactivation of SP at a SP synapse. If we assume that the affinity of SP for its receptor must be, as for many other neuroactive peptides, in the nanomolar range, and therefore that the concentration of SP in the synaptic cleft ought not to exceed micromolar concentrations, then the degradation of SP should be carried out by membrane-bound enzymes presenting either of the following properties: low K_m value (in the order of 10^{-7}–10^{-6} M), high local concentration or high k_{cat} value. Since efficient proteolytic enzymes generally display k_{cat} values lower than 100 sec^{-1}, one must assume that the degradation of SP takes place at synapses relatively rich in proteases showing high affinities for the substrate SP. The lower the K_m value, the smaller should be the effective concentration of these degrading enzymes at the synapse. Soluble peptidases found in the brain to degrade SP with K_m values around 10^{-3} M (12) are probably not the specific enzymes inactivating SP at SP synapses.

With these considerations in mind, we decided to study the hydrolysis of SP analogues by synaptosomal fractions of rat brain using peptides at a concentration of $\sim 10^{-7}$ M. The hydrolysis was followed with time by removing an aliquot of the incubation mixture and measuring the decrease in its ability to contract a guinea pig ileum.

Synaptosomal fractions were found to hydrolyse SP analogues in a very efficient way. Figure 2 illustrates the hydrolysis of the C-terminal heptapeptide carried out by these preparations at room temperature. The half-life of the peptide can be calculated as the time needed to decrease the contraction from 80% to 50% (figure 1 shows that in the linear range of the dose-response curve, doubling of the ED_{50} concentration gives a contraction of 80%). The half-life is 2 min when the peptide is incubated with synaptosomes at 1 mg/ml.

This concentration of protein was used since most binding studies are carried out under such concentration conditions. Because of the fast rate of degradation of the C-terminal heptapeptide, one ought to question the feasibility of

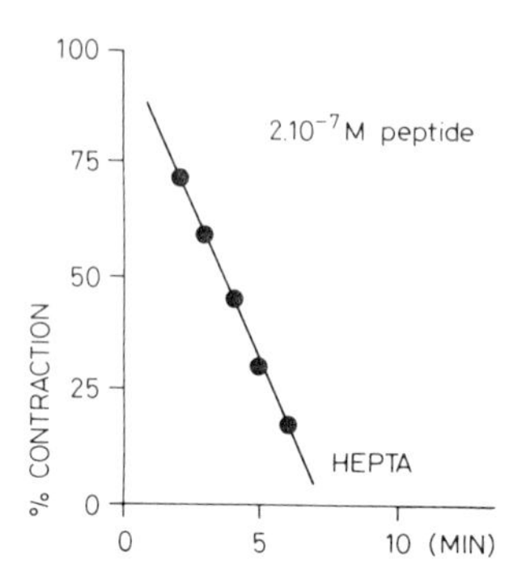

Fig. 2. Hydrolysis of 2×10^{-7} M C-terminal heptapeptide by 1 mg/ml crude synaptosomal fraction of rat brain at room temperature. The crude synaptosomal fraction was prepared as follows: Rat brains were homogenized 10:1 (v/w) at 1,400 rpm using 8 up and down strokes of a Teflon pestle in a glass homogenizer. The buffer used was an ice-cold solution of 0.32 M sucrose, 20 mM sodium phosphate, pH 7.4. After centrifugation at 1,000 g for 10 min, the pellet was discarded and the supernatant centrifuged at 20,000 g for 20 min. The pellet was washed twice by resuspension and recentrifugation. It was finally resuspended in a buffer containing 136 mM NaCl, 2.7 mM KCl, 1.47 mM KH_2PO_4, 1 mM EGTA, 1 mM dithiothreitol, at pH 7.4.

equilibrium binding studies using this ligand as a probe of the SP receptor. If we assume that the SP receptor concentration in synaptosomal fractions ranges from 10 to 100 fmole/mg, one can easily calculate the time needed by a ligand to reach binding equilibrium.

Assuming a second-order association rate constant $k_1 = 10^6$ M^{-1} sec^{-1} (similar to that of neurotensin (13)), a ligand concentration of 10^{-9} M, and 10^{-10} M receptor sites, one readily calculates that about 12 min are needed to reach half occupancy of the receptor sites. It is clear therefore that ligands as sensitive to degradation as the C-terminal heptapeptide cannot be used in equilibrium binding studies to probe the SP receptor. Such studies may be feasible only if one can lower the sensitivity of the peptide to proteolytic degradation. This might be carried out for instance by introducing substitute amino acids in the original sequence of SP in order to stabilize the molecule. Such examples of peptides are shown in table II. Assuming that the peptide linkages near glycine^9 were particularly prone to hydrolysis, we have substituted this residue with D-alanine or β-alanine. However, as seen in table II, these substitutions markedly decrease the contracting efficacy of the peptides and therefore render them inadequate for binding studies.

Since the natural sequence of the C-terminal heptapeptide of SP seemed the most appropriate for a high affinity interaction with its receptor, we then considered the possibility of a stabilization of this molecule by either of the two

Table III. Effect of inhibitors on the hydrolysis of 4×10^{-7} M heptapeptide by crude synaptosomal preparation of rat brain

	Half-life of peptide
None	3 min
Leupeptin 20 μg/ml	3
Chemostatin 20 μg/ml	3
Antipain 20 μg/ml	5
Bacitracin 10 μg/ml	3
Phenylmethylsufonyl fluoride 10^{-3} M	3
o-Phenanthroline 10^{-3} M	12

following ways: 1) A protection of the peptide from degradation might be achieved by using appropriate inhibitors of proteases. 2) Since the C-terminal residue of SP is protected by an amino group, one may assume that the latter prevents a degradation of the peptide by carboxypeptidases. However, SP and SP partial sequences possess a free N-terminal and therefore protection of this amino group might hinder the action of aminopeptidases.

Using the first approach, we screened several known protease inhibitors for their ability to decrease the hydrolysis of the free C-terminal heptapeptide of SP by synaptosomal preparations.

Table III lists some of these inhibitors together with the extent of protection that was obtained. None of these inhibitors was found to afford total protection and this result indicated that several proteases are probably involved in the degradation of the peptide.

Leu-MetNH$_2$, an inhibitor of degradation of SP by epithelial cells (11), was found to fully inhibit the degradation of SP by rat brain homogenate (fig. 3). However the protection by Leu-MetNH$_2$ was only transient and lasted until the dipeptide itself was degraded.

In the second approach, we studied the effects resulting from the blocking of the free amino group of the peptide on its degradation by synaptosomal fractions. Whereas the C-terminal heptapeptide was rapidly degraded by synaptosomes at a concentration of 1 mg/ml, the blocked derivative showed a marked resistance to degradation and therefore the protein concentration had to be increased to 5 mg/ml to observe a measurable degradation (fig. 4). The most direct conclusion from this experiment is that two families of proteases, aminopeptidases and endopeptidases, are involved in the degradation of the heptapeptide. To further strengthen this conclusion, we investigated the effect of metal complexing agents on the degradation, taking account of the fact that

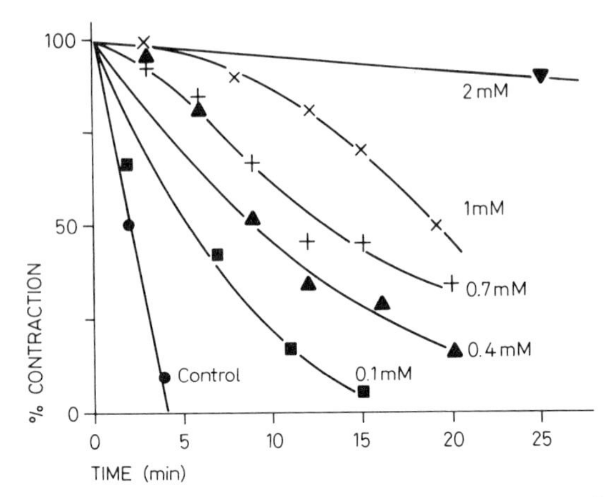

Fig. 3. Inhibition Leu-Met-NH_2 of the hydrolysis of 10^{-7} M SP by 1% rat brain homogenate. Rat brain homogenate was prepared from male Sprague-Dawley rats of 50 days. The homogenization buffer contained 136 mM NaCl, 2.7 mM KCl, 1.47 mM KH_2 PO_4, 1 mM EGTA, 1 mM dithiothreitol, at pH 7.4. The homogenate was filtered through cheese cloth and diluted to a final concentration of 1%.

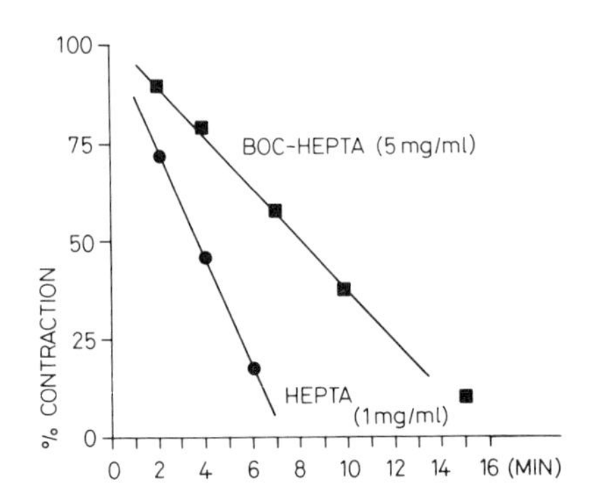

Fig. 4. Hydrolysis of the N-terminal free (•) and blocked (▪) heptapeptide by synaptosomes. Both peptides were at 2 × 10^{-7} M.

many aminopeptidases are metal-requiring enzymes whereas most endopeptidases are not. As shown in figure 5A and 5B, we indeed found that *o*-phenanthroline, a metal complexing agent, markedly slows down the degradation of the free hexapeptide but has no effect on the blocked derivative.

Table IV includes data from a systematic study on the degradation of various partial sequences of SP in their free and blocked forms. Inspection of the results clearly indicates the high sensitivity of the free peptides to aminopepti-

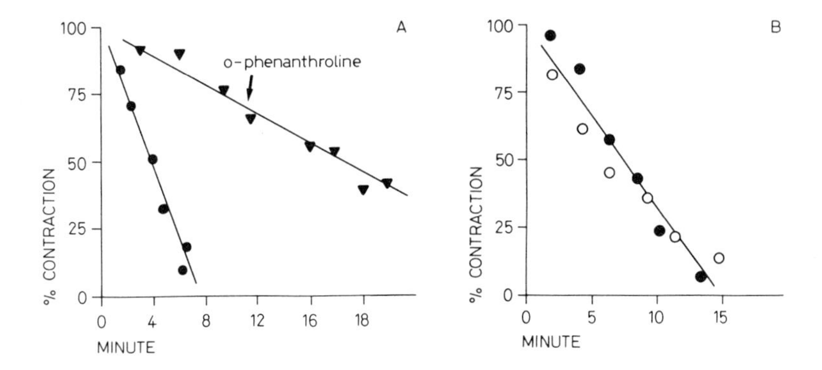

Fig. 5. A. Hydrolysis of 2×10^{-7} M hexapeptide by synaptosomes in the presence (▾) or absence (•) of 10^{-3} M *o*-phenanthroline. B. Hydrolysis of 2×10^{-7} M blocked hexapeptide by synaptosomes in the presence (♦) or absence (•) of 10^{-3} M *o*-phenanthroline.

Table IV. Half-lives (min) of Substance P analogues in the presence of rat brain synaptosomal preparation

Peptide (2×10^{-7} M)	Synaptosomal concentration	
	1 mg/ml free	5 mg/ml blocked
(6–11) Hexa	1.5 ± 0.5	2.5 ± 1
(5–11) Hepta	2 ± 0.5	5 ± 2
(4–11) Octa	5 ± 2	6 ± 2
Substance P	5 ± 2	—

dases. However, significant differences are observed from peptide to peptide. The free octapeptide and SP are clearly more resistant to degradation than the shorter hepta- or hexapeptide and it therefore appears that the presence of the proline[4] contributes to the stabilization of the molecule. Since the rate of hydrolysis of the free octapeptide is virtually the same as that of SP, one must assume that SP is not directly degraded by aminopeptidases. Indeed if this had been the case, the time needed to remove the four N-terminal residues of SP should have been longer than that needed to cleave the ultimate proline residue of the octapeptide.

We therefore suggest that the degradation of SP is carried out in two steps: SP is first cleaved by a post-proline cleaving enzyme splitting the molecule into two peptides, an N-terminal tetrapeptide and a C-terminal heptapeptide, the latter being degraded, in a second step, by aminopeptidases. Recently evidence has been presented for the existence in brain of a soluble post-proline cleaving enzyme for which we have found SP to be a very good substrate (4). This enzyme splits SP into the two above mentioned fragments and displays for this substrate a very high affinity of 10^{-6} M (4). This value is remarkably high, particularly if one takes into consideration a catalytic constant k_{cat} ~50 sec^{-1} that was found for such an enzyme from kidney using a variety of synthetic substrates (21). The k_{cat}/K_m value of this enzyme is therefore in the range of 10^7–10^8 M^{-1} sec^{-1} and is comparable to that found for another neurotransmitter-degrading enzyme: acetylcholinesterase.

The enzyme is soluble but preliminary experiments indicate its presence in synaptosomal fractions as well. It remains to be established whether this post-proline cleaving enzyme is indeed responsible for the specific degradation of SP at SP synapses. Until such evidence is provided, one can nevertheless attempt to assess the properties of this enzyme in the light of those expected from a specific SP-inactivating mechanism. Assuming Michaelian kinetics and considering both the half-life of the SP substrate at low concentration and the k_{cat}/K_m value mentioned above, one can calculate the concentration of the enzyme to be around 20 pmole/mg protein. This is a reasonable number particularly if one assumes the SP receptor to be present in these synaptosomal preparations at a concentration of 10–100 fmole/mg. If the affinity of the receptor for SP is in the nanomolar range and if SP is released in the synaptic cleft at a local concentration of 10^{-7}–10^{-6} M, then the ratio of concentration between the SP-degrading enzyme and the SP receptor is such that competition could take place, and SP would have equal probability to elicit a postsynaptic response and to be degraded.

Summary

A series of C-terminal partial sequences of Substance P, either free or blocked at the amino terminus, has been synthesized. The peptides were examined for their relative potencies as smooth muscle contracting agents and for their rates of degradation by rat brain synaptosomes. The C-terminal hexapeptide in both the free and blocked forms displays activity comparable to that of the longer C-terminal peptides as well as to that of the native undecapeptide. The blocked peptides, however, are much more stable than the corresponding free peptides. Among the free peptides Substance P is degraded virtually at the same rate as the C-terminal octapeptide but significantly more slowly than the C-terminal hexa- or heptapeptides. These patterns of inactivation together with the response to inhibition by various protease inhibitors indicate that both endopeptidase(s) and aminopeptidase(s) are involved in the degradation of Substance P. Degradation of the Substance P molecule at Substance P synapses may occur in two steps, first by a specific endopeptidase splitting the molecule into an N-terminal tetrapeptide and a C-terminal heptapeptide, the latter being degraded in a second step by aminopeptidases.

Acknowledgements

We thank Drs. *D. Michaelson* and *M. Sokolovsky* for the use of their Physiograph Apparatus. We are indebted to *Susan M. Price* for excellent laboratory assistance.

This research was supported by grants from the Délégation Générale à la Recherche Scientifique et Technique and from the Israel Commission for Basic Research. *S.B.* is the incumbent of the M & W Levine Career Development Chair.

References

1 Akopyan, T.N.; Arutunyan, A.A.; Lajtha, A., and Galoyan, A.A.: Acid proteinase of hypothalamus, purification, some properties, action on somatostatin and Substance P. Neurochem. Res. *3:* 89–99 (1978).

2 Benuck, M.; Grynbaum, A., and Marks, N.: Breakdown of somatostatin and Substance P by cathepsin D purified from calf brain by affinity chromatography. Brain Res. *143:* 181–185 (1977).

3 Benuck, M. and Marks, N.: Enzymatic inactivation of Substance P by a partially purified enzyme from rat brain. Biochem. biophys. Res. Commun. *65:* 153–160 (1975).

4 Blumberg, S.; Teichberg, V.I.; Charli, J.L.; Hersh, L.B., and McKelvy, J.F.: Cleavage of Substance P to an N-terminal tetrapeptide and a C-terminal heptapeptide by a post-proline cleaving enzyme from bovine brain. Proc. Fed. Am. Socs exp. Biol. *38:* 350 (1979).

5 Bury, R.W. and Mashford, M.L.: Biological activity of C-terminal partial sequences of Substance P. J. med. Chem. *19:* 854–856 (1976).

6a Chang, M.M.; Leeman, S.E., and Niall, M.D.: Amino acid sequence of Substance P. Nature new Biol. *232:* 86–87 (1971).

6b Studer, R.O.; Trzeciak, H. und Lergier, N.: Isolierung und Aminosaüresequenz von Substanz P aus Pferdedam. Helv. chim. Acta *580:* 860–866 (1973).

7 Euler, U.S. von and Gaddum, J.H.: An unidentified depressor substance in certain tissue extracts. J. Physiol., Lond. *72:* 74–87 (1931).

8 Gullbring, B.: Inactivation of Substance P by tissue extracts. Acta physiol. scand. *6:* 246–255 (1943).

9 Henry, J.L.: Substance P and pain: a possible relation in afferent transmission; in von Euler and Pernow, Substance P; pp. 231–240 (Raven Press, New York 1977).

10 Hökfelt, T.; Johansson, O.; Kellert, J.-O.; Ljungdahl, A.; Nilsson, G.; Nygards, A., and Pernow, B.: Immunohistochemical distribution of Substance P; in von Euler and Pernow, Substance P; pp. 117–145 (Raven Press, New York 1977).

11 Johnson, A.R. and Erdös, E.G.: Inactivation of Substance P by cultured human endothelial cells; in von Euler and Pernow, Substance P; pp. 253–260 (Raven Press, New York 1977).

12 Kato, T.; Nagatsu, T.; Fukasawa, K.; Harada, M.; Nagatsu, I., and Sakakibara, S.: Successive cleavage of N-terminal Arg Pro and Lys Pro from Substance P but no release of Arg Pro from bradykinin by X-Pro dipeptidyl aminopeptidase. Biochem. biophys. Acta *525:* 417–422 (1978).

13 Kitabgi, P.; Carraway, R.; van Rietschoten, J.; Granier, C.; Morgat, J.L.; Menez, A.; Leeman, S., and Freychet, P.: Neurotensin: Specific binding to synaptic membranes from rat brain. Proc. natn. Acad. Sci. USA *74:* 1846–1850 (1977).

14 Konishi, S. and Otsuka, M.: The effects of Substance P and other peptides on spinal neurons of the frog. Brain Res. *65:* 397–410 (1974).

15 Lembeck, F.: Zur Frage der zentralen Übertragung afferenter Impulse. III. Mitteilung. Das Vorkommen und die Bedeutung der Substanz P in den dorsalen Wurzeln des Rückenmarks. Naunyn-Schmiedebergs Arch. exp. Path. Pharmak. *219:* 197–213 (1953).

16 Nakata, Y.; Kusaka, Y.; Segawa, T.; Yajima, H., and Kitagawa, K.: Substance P: Regional distribution and specific binding to synaptic membranes in rabbit central nervous system. Life Sci. *22:* 259–268 (1978).

17 Otsuka, M. and Konishi, S.: Release of Substance P like immunoreactivity from isolated spinal cord of newborn rat. Nature, Lond. *264:* 83–84 (1976).

18 Otsuka, M. and Takahashi, T.: Putative peptide neurotransmitters. Ann. Rev. Pharmacol. Toxicol. *17:* 425–439 (1977).

19 Pelletier, G.; Leclerc, R., and Dupont, A.: Electron microscope immunohistochemical localization of Substance P in the central nervous system of the rat. J. Histochem. Cytochem. *25:* 1373–1380 (1977).

20 Rosell, S.; Bjorkroth, U.; Chang, E.; Yamaguchi, I.; Wan, Y.P.; Rackur, G.; Fischer, G., and Folkers, K.: Effects of Substance P and analogs on isolated guinea pig ileum; in von Euler and Pernow, Substance P; pp. 83–88 (Raven Press, New York 1977).

21 Walter, R. und Yoshimoto, T.: Postproline cleaving enzyme: Kinetic studies of size and stereospecificity of its active site. Biochemistry *17:* 4139–4144 (1978).

22 Yanaihara, N.; Yanaihara, C.; Hirohashi, M.; Sato, H.; Tizuka, Y.; Hashimoto, T., and Sakagami, M.: Substance P analogs: synthesis, and biological and immunological properties; in von Euler and Pernow, Substance P; pp. 27–33 (Raven Press, New York 1977).

Dr. V.I. Teichberg, Department of Neurobiology,
The Weizmann Institute of Science, Rehovot (Israel)

Prog. biochem. Pharmacol., vol. 16, pp. 95–108 (Karger, Basel 1980)

Studies on the Properties of Benzodiazepine-Binding Sites from Calf Cortex[1]

Yadin Dudai[2] *and Rivka Sherman-Gold*

Department of Neurobiology, The Weizmann Institute of Science, Rehovot

Introduction

Mammalian brain has recently been shown to contain high affinity binding sites for benzodiazepines (1, 2, 14, 15, 20). These sites, which appear to be associated with GABAergic mechanisms (3, 5, 8, 22), presumably mediate the potent pharmacological action of the benzodiazepines *in vivo* (17, 23).

Studies on the molecular properties of benzodiazepine-binding sites may shed light on mechanisms by which the brain modulates transmitter efficacy and regulates behavior, as well as on the nature of molecular lesions underlying neurological and affective disorders. Properties of benzodiazepine-binding sites have been studied until now in rat brain using mainly [^{3}H] diazepam. Experiments were performed at 0–4°C because very little specific binding of [^{3}H] diazepam was detected at physiological temperatures (2, 15). Calf brain can serve as a suitable tissue for biochemical investigation of benzodiazepine receptors since it provides a rich source of these sites (16). In the following, we describe properties of particulate benzodiazepine-binding sites from calf cortex, as revealed by specific binding of [^{3}H] flunitrazepam ([^{3}H] FNZ). The use of this high affinity benzodiazepine makes it possible to study the putative receptor at physiological temperatures. Studies on the properties of detergent-treated benzodiazepine receptors from calf cortex are now in progress (*Sherman-Gold and Dudai,* in preparation).

Materials and Methods

Preparation of Tissue. Calf brain was obtained from a local slaughterhouse and frozen at −80° C. The brain was thawed and the entire cortex removed, cut into cubes and homogenized (20% w/v) in a Sorvall Omni-Mix (speed 6) for 15 sec in ice-cold 0.32 M sucrose. The

[1] The excellent technical assistance of *Shoshana Nahum* is gratefully acknowledged.

[2] Incumbent of the Barecha Foundation Career Development Chair.

homogenate was centrifuged at 650 × g for 10 min. The pellet was discarded and the supernatant collected and frozen in 10–20 ml aliquots at –80°C. For preparation of the homogenate used in binding studies (which will be referred to as H), aliquots were thawed and homogenized in ice-cold 0.32 M sucrose (1:1 v/v) in a glass-Teflon homogenizer. For preparation of crude membranes fraction (which will be referred to as M), the following procedure was employed: H was centrifuged at 30,000 × g for 40 min. The supernatant was removed and stored at –20°C and the pellet was rehomogenized in double-distilled water, frozen for approximately 2 h at –20°C, thawed, and centrifuged again at 30,000 × g for 40 min. The resulting pellet was homogenized in 0.12 M NaCl, 0.05 M Tris, pH 7.4, and centrifuged again. The latter procedure was repeated and the resulting pellet was homogenized in the original homogenate volume, to yield the washed membranes.

Chemicals. [^{3}H]flunitrazepam ([^{3}H]FNZ), 84.3 Ci/mmol, was obtained from New England Nuclear (Boston, MA). Flunitrazepam and RO-5-4864 were the gift of Dr. *H. Möhler,* Basel. Diazepam and medazepam were donated by Assia Ltd., Jerusalem, and nitrazepam, flurazepam and chlorodiazepoxide were donated by Ikapharm Ltd., Kfar Saba. GABA and bicuculline were purchased from Sigma (St. Louis, MO) and (±)-trans-3-aminocyclopentane carboxylic acid was provided by Dr. *M. Segal.* Sodium deoxycholate was from Schuchardt, München. Other chemicals were of analytical grade or (in the case of some detergents) of the highest technical grade available.

[^{3}H]FNZ Binding Assay. Several assay methods were tested (see under *Results*). Unless otherwise indicated, the following routine assay was employed: Aliquots of homogenate or washed membranes preparation were incubated at the appropriate temperature in 0.12 M NaCl, 0.05 M Tris-Cl, pH 7.4 (=buffer), in final volume 70–400 μl. Reaction was started by addition of [^{3}H]FNZ and terminated after appropriate incubation by diluting with 2 ml of ice-cold buffer, followed immediately by vacuum filtration through a glass-fiber filter (GF/C, Tamar, Israel). The filter was then rapidly washed 3 times with 2 ml portions of ice-cold buffer, dried, placed in vials containing 5 ml of 33% (v/v) Triton X-100, 0.8% PPO and 0.01% POPOP in toluene, and counted by liquid scintillation spectrometry. Specific binding of [^{3}H]FNZ was defined as total binding minus the binding occurring in the presence of 10^{-5} M diazepam.

Protein was determined according to *Lowry et al.* (11), using BSA as a standard.

Results

Binding of [^{3}H]FNZ to Calf Cortex Preparations

Preliminary experiments were conducted to test the suitability and efficiency of various potential assays of [^{3}H]FNZ binding to calf cortex preparations. The assays included filtration through Millipore (EGWP, HAWP, GSWP and PHWP), glass-fiber and DEAE-cellulose filters, centrifugation and ammonium sulfate precipitation (various concentrations at a range of 30–70% saturation were tested, to reveal putative soluble sites).

Of all the filters tested, glass-fiber filters gave optimal results with minimal non-specific adsorption of [^{3}H]FNZ to the filter at all the temperatures tested (0–37°C). When the reaction was rapidly terminated with an ice-cold buffer, filtration assay gave essentially the same results as centrifugation assay, indicating negligible off-reaction under these conditions (see below). Employing glass-

Table I. Subcellular distribution of [^{3}H] FNZ-binding sites in calf cortex homogenate

Assay method:	*Buffer*		*Buffer*	*1% Triton X-100*
Activity	*I*		*II*	*II*
	pmol	*pmol/mg protein*	*pmol*	*pmol*
Fraction				
$S_{500 \times g}$	84.3	1.27	71.8	81.0
$S_{20,000 \times g}$	0.6	0.04	0.7	15.8
$P_{20,000 \times g}$	67.0	1.66	74.7	53.9
$S_{100,000 \times g}$	0.1	0.01	0.4	10.7
$P_{100,000 \times g}$	0.2	0.38	1.2	8.2

An aliquot of 10 ml of a homogenate preparation ($S_{500 \times g}$ prepared as described under *Methods* in 0.32 M sucrose) was divided into two portions. 5 ml of 0.12 M NaCl, 0.05 M Tris-Cl, pH 7.4 were added to one of the portions whereas 5 ml of 1% Triton X-100 in the same buffer were added to the other portion. The samples were rehomogenized in a glass-Teflon homogenizer and centrifuged for 30 min at 20,000 × g, to yield $S_{20,000 \times g}$ and $P_{20,000 \times g}$. Further centrifugation for 60 min at 100,000 × g yielded $S_{100,000 \times g}$ and $P_{100,000 \times g}$. [^{3}H] FNZ binding was determined at 0°C (at [^{3}H] FNZ concentration of 10 nM) by the standard glass-fiber method (I) or by precipitation with cold 40% saturated, neutralized ammonium sulfate followed by filtration on glass-fiber filters (II).

fiber filters (GF/C), the binding observed with a given concentration of [^{3}H] FNZ was linearly proportional to the amount of tissue present at the range employed (50–300 μg protein). Essentially all the binding was located at the fraction sedimenting between 500 × g and 20,000 × g (table I). The subcellular sedimentation profile was not altered when an ammonium sulfate precipitation assay was used although the latter assay did reveal specific [^{3}H] FNZ-binding sites at high-speed supernatant fractions of a detergent-treated homogenate (tables I, II). It was therefore concluded that essentially all [^{3}H] FNZ-binding sites in a homogenate with no detergent present are particulate.

Treatment with various detergents markedly altered the subcellular distribution of the benzodiazepine-binding sites and released a substantial part of the binding activity into a high-speed supernatant (table II). Sodium deoxycholate (DOC) was the most efficient. About 10% of the DOC-treated binding sites remained in the supernatant even after centrifugation at 200,000 × g for 2 h. The physical properties of the detergent-treated sites are now under investigation. Unless otherwise indicated, the results presented in this work refer to the particulate binding sites in a non-detergent homogenate.

Table II. Effect of detergents on the subcellular distribution of [^{3}H]FNZ-binding sites in calf cortex

Medium	M	$P_{100,000 \times g}$	$S_{100,000 \times g}$
Buffer	100	103	0
1% Triton X-100	70	76	13
1% Tween 80	74	103	0
1% Digitonin	92	66	3
0.5% Tauro DOC	57	32	20
0.1% DOC	90	80	0
0.25% DOC	69	48	28
0.5% DOC	118	37	35
1% DOC	76	44	32
1% Taurocholate	79	76	27
1% Cholate	62	55	3
1% Brij 35	114	124	0
1% NP40	103	64	22
0.1% SDS	102	85	2
2 M NaCl	100	87	0

Washed membranes preparations (= M) prepared in buffer (0.06 M NaCl, 0.025 M Tris-Cl, pH 7.4) as described under *Methods,* were incubated with stirring for 2 h at 4°C in the presence of buffer or of buffer + the appropriate detergent. The preparations were then centrifuged for 1 h at 100,000 × g to yield the supernatant ($S_{100,000 \times g}$) and the pellet ($P_{100,000 \times g}$). The latter was dispersed, and specific [^{3}H]FNZ binding was determined in aliquots of the membranes and of the resulting pellet and supernatant by the ammonium sulfate assay. Triton X-100 and NP40 gave variable and sometimes very high blanks in this assay. The results are expressed as percent of total specific [^{3}H]FNZ binding in each fraction, where 100% is the activity of the membranes incubated under the experimental conditions in the presence of buffer alone. DOC = Sodium deoxycholate; SDS = sodium dodecyl sulfate.

Chloride ions were found to be required for optimal binding, but no requirement for specific cations was revealed (table III). The optimal NaCl concentration was found to be ~0.1 M. Phosphate ions of an equivalent ionic strength were less effective than Cl^-. These results are in agreement with a recent report by *Costa et al.* (4) for rat brain but contradict previous results of *Bosman et al.* (1), *Mackerer and Kochman* (12) and *Möhler and Okada* (15). Hg^{+2} at concentrations higher than 1 mM was inhibitory for binding (table III). Dithiothreitol, N-ethylmaleimide and EDTA had no significant effect. Binding was essentially the same at a pH range of 6.5–9.5 and was not dependent upon the buffer used (citrate, phosphate or Tris). The routine assay conditions described under *Methods* were chosen on the basis of the above findings.

Table III. Effect of ions on [^{3}H] FNZ binding to calf cortex washed membranes preparation

Medium	*[^{3}H] FNZ bound, fmol*
1 mM Tris-Cl, pH 7.4	13.4
+ 1 mM NaCl	13.7
+ 10 mM NaCl	19.0
+ 100 mM NaCl	29.6
+ 1 M NaCl	27.5
+ 1 mM $CaCl_2$	17.1
+ 33 mM $CaCl_2$	24.1
+ 1 mM $MgCl_2$	17.0
+ 33 mM $MgCl_2$	23.4
20 mM Tris-Cl, pH 7.4	19.2
+ 10 mM $HgCl_2$	7.2

Washed membranes preparation was prepared as described under *Methods* except that the pellet obtained in each preparative step was dispersed in double-distilled water instead of buffer. The reaction mixture contained 45 μg membrane protein and 5 nM [^{3}H] FNZ. Incubation was for 60 min at 0°C.

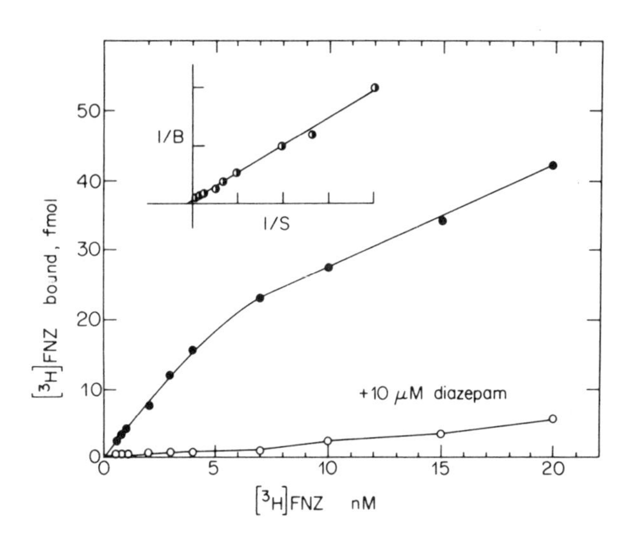

Fig. 1. The level of binding of [^{3}H] FNZ to calf cortex washed membranes preparation after incubation for 30 min at 37°C in the presence of various concentrations of ligand. •——•, Specific binding, i.e. total binding minus binding in the presence of 10^{-5} M diazepam; o——o, non-specific binding, i.e. binding in the presence of 10^{-5} M diazepam. Insert: double reciprocal plot of specific binding. Each reaction mixture contained 80 μg membrane protein.

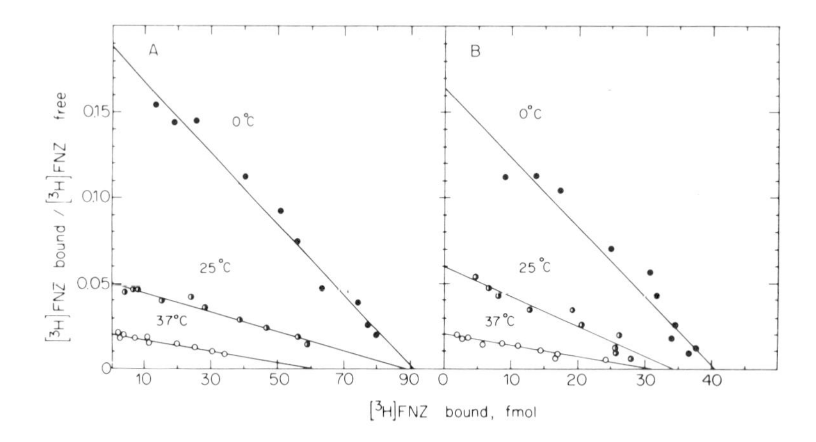

Fig. 2. Specific binding of [^{3}H] FNZ to calf cortex washed membranes preparation (A) and to homogenate preparation (B) as a function of incubation temperature. Each reaction mixture contained 80 μg protein in (A) and 60 μg protein in (B). Data are presented as *Scatchard* plots.

Table IV. Affinity for [^{3}H] FNZ (K_D, nM) of benzodiazepine-binding sites from calf cortex under various conditions

Preparation	M	M + sup	M + GABA	H	H + GABA	S(DOC)	S(DOC) + GABA
Temperature							
0°C	1.9 ± 0.2	0.9	1.1 ± 0.1	1.2 ± 0.2	0.9	1.6 ± 0.1	1.1 ± 0.2
25°C	9.0	–	–	2.9 ± 0.1	–	–	–
37°C	14.8 ± 1.0	7.7	7.7	7.6 ± 0.3	8.1	–	–

Apparent dissociation constants were determined from *Scatchard* plots. [^{3}H] FNZ concentrations used were 0.5–20 nM, and reaction was carried out as described under *Methods.* M and H are washed membranes and homogenates, respectively, prepared as described under *Methods*; sup is the supernatant obtained after 100,000 × g centrifugation for 1 h of a homogenate in buffer with no detergent (see also fig. 3). S(DOC) is the supernatant obtained after 100,000 × g centrifugation for 1 h of washed membranes treated with 0.5% sodium deoxycholate (see also table II). GABA concentration was 10^{-5} M. Values for multiple determinations are mean ± SEM.

Binding to Washed Membranes Preparations

Specific binding of [^{3}H] FNZ to washed membranes (M) preparations, defined as total binding minus binding in the presence of 10^{-5} M diazepam, was saturable (fig. 1). Non-specific binding at the [^{3}H] FNZ concentrations employed (0.5–20 nM) was low at all temperatures tested (e.g., fig. 1). The apparent dissociation constant calculated from binding isotherms was temperature dependent, and affinity decreased with increasing temperature (fig. 2, table IV). Low ionic strength ($I < 0.01$) did not have a significant effect on affinity, but treating the membranes with Triton did seem to decrease the affinity of the remaining membrane-bound sites when tested at 0°C ($K_D = 3.50 \pm 0.3$). Binding isotherms revealed only a single class of binding sites at the concentration range tested (fig. 2), and no cooperativity was revealed either at 0°C (Hill coefficient = 1.01) or at 37°C (Hill coefficient = 0.99).

Binding to Homogenate Preparations

Specific binding of [^{3}H]FNZ to homogenate (H) preparations was also saturable with a single class of binding sites and the K_D was again temperature dependent (table IV). However, apparent affinity was higher than that revealed in washed membrane preparations (fig. 2, table IV).

Modulation of Washed-Membranes Sites

The difference between the apparent affinity of H and M preparations suggested that a factor removed from the membranes during repeated washings increases the affinity of their sites. Indeed, addition of aliquots of a 100,000 × g supernatant to the membranes increased [^{3}H]FNZ binding by increasing affinity (table IV) without significantly altering the number of binding sites. The modulation was already observed at a dilution of 1:200 of the supernatant (fig. 3), and was not altered by boiling or treating the supernatant with 1% w/w pronase for 7 h at 37°C.

The effect of the supernatant could be mimicked by GABA (figs. 3, 4) and the GABAergic agonist (±)-trans-3-aminocyclopentane carboxylic acid (18). β-Alanine had a much smaller effect (data not shown). The increase in affinity caused by GABA could be observed both at 0°C and at 37°C (fig. 4). Half maximal activity was reached at a GABA concentration of about 10^{-6} M. Hill coefficient for GABA modulation was found to be $n_H = 0.91 \pm 0.06$ ($n = 3$), indicating probably no cooperativity. Chloride ions were not found to be obligatory for GABA modulation.

The effect of the 100,000 × g supernatant and of GABA were antagonized by bicuculline but not by glutamate (10^{-4} M), amobarbital (10^{-4} M), Hg^{+2} (1 mM), ConA (0.5 mg/ml), wax-bean agglutinin (0.1 mg/ml) or wheat-germ agglutinin (0.2 mg/ml). Thus, in a typical experiment [^{3}H]FNZ binding to a washed membranes preparation (5 nM [^{3}H]FNZ, 37°C) was enhanced 1.82-fold by 10^{-5} M GABA and 1.73-fold by a 1:40 dilution of 100,000 × g supernatant, but the corresponding binding levels were only 0.91 and 0.99 in the presence of 10^{-4} M bicuculline. The latter had only a small effect (<15%) on [^{3}H]FNZ binding under these conditions. Whereas the effect of 10^{-4} M bicuculline on [^{3}H]FNZ binding to washed membranes was small, it caused a decrease of about 40% of [^{3}H]FNZ binding to homogenate preparations. This is compatible with

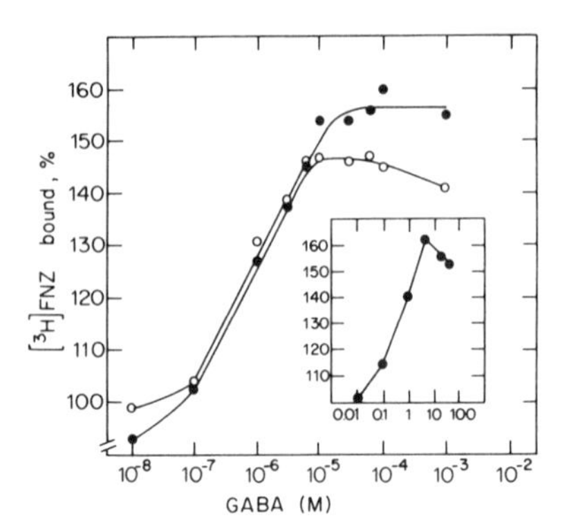

Fig. 3. Effect of GABA on the binding of $[^3H]$FNZ to calf cortex washed membranes preparation. Reaction was carried out at 37°C for 30 min and $[^3H]$FNZ concentration was 5 nM (●——●) or 15 nM (○——○). The insert shows, for comparison, the effect on binding of aliquots of 100,000 × g supernatant (μl in a final volume of 100 μl).

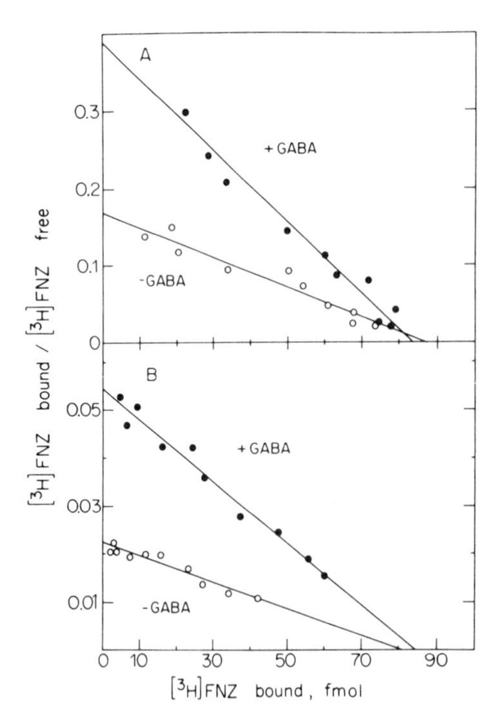

Fig. 4. Effect of 10^{-5} M GABA on specific binding of $[^3H]$FNZ to calf cortex washed membranes preparation, measured at 0°C (A) and at 37°C (B). Data are presented as *Scatchard* plots.

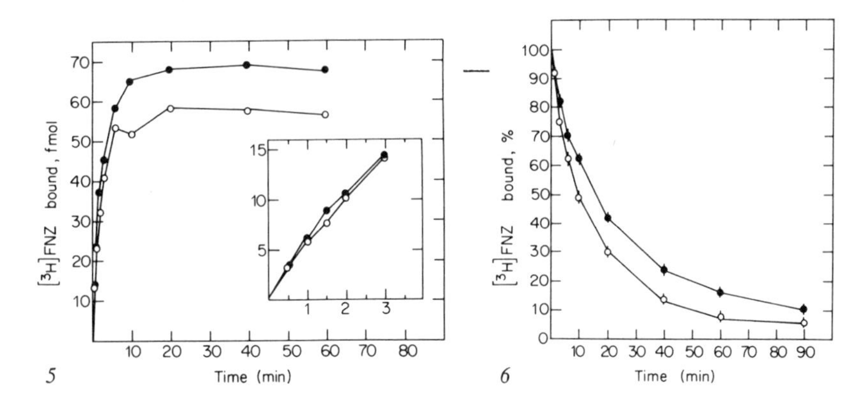

Fig. 5. Binding of [^{3}H]FNZ (final concentration 5 nM) to calf cortex washed membranes preparation at 0°C as a function of time, in the presence (●——●) and in the absence (○——○) of 10^{-5} M GABA. Insert: initial rate determined under similar conditions but with 0.5 nM (^{3}H]FNZ.

Fig. 6. Dissociation of [^{3}H]FNZ from calf cortex washed membranes preparation as a function of time, in the presence (●——●) and in the absence (○——○) of 10^{-5} M GABA. The reaction was started by addition of [^{3}H]FNZ (final concentration 5 nM) and was carried out at 0°C. After >40 min, non-labeled flunitrazepam was added to a final concentration of 10^{-4} M, and aliquots were assayed at various times later as described under *Methods*. Each point represents mean ± SEM for three experiments.

the observation that binding activity observed in the homogenate is already modulated by endogenous GABAergic ligand(s). Indeed, addition of 10^{-5} M GABA to homogenate preparations did not cause further significant increase in affinity (table IV).

The effect of GABA on the on- and off-reaction of [^{3}H]FNZ with the washed membranes preparation was investigated. Under the conditions employed 10^{-5} M GABA had only a small effect on the on-reaction (fig. 5), but markedly slowed the off-reaction (fig. 6). Thus, assuming the simplest kinetic model for receptor (R)-ligand (L) interaction, i.e.

$$R + L \underset{k_{-1}}{\overset{k_1}{\rightleftharpoons}} R \cdot L \ ,$$

k_1, the on-rate constant at 0°C was found to be $(9.6 \pm 0.6) \times 10^5 M^{-1} s^{-1}$ in the absence of GABA and $(10.0 \pm 1.0) \times 10^5 M^{-1} s^{-1}$ in the presence of 10^{-5} M GABA (n = 5 each). However, k_{-1}, the off-rate constant, was found to be $(11.9 \pm 0.9) \times 10^{-4} s^{-1}$ in the absence of GABA but $(6.9 \pm 0.2) \times 10^{-4} s^{-1}$ in the presence of 10^{-5} M GABA (n = 3 each). The apparent dissociation constants were hence calculated to be $K_D = 1.2 \times 10^{-9}$ M in the absence of GABA and 0.7×10^{-9} M

Table V. Effect of various ligands on [^{3}H]FNZ binding to calf cortex benzodiazepine-binding sites

Preparation	M	H	M	H
Temperature	0°C	0°C	37°C	37°C
Ligand				
flunitrazepam	1.6×10^{-9}	–	4.5×10^{-8}	
diazepam	2.0×10^{-8}	2.3×10^{-9}	1.8×10^{-7}	1.0×10^{-7}
nitrazepam	1.6×10^{-8}	1.0×10^{-8}	1.4×10^{-7}	7.0×10^{-8}
flurazepam	4.6×10^{-8}	1.8×10^{-8}	3.0×10^{-7}	1.4×10^{-7}
chlorodiazepoxide	5.5×10^{-7}	2.7×10^{-7}	2.1×10^{-6}	1.1×10^{-6}
medazepam	7.4×10^{-6}	1.5×10^{-6}	1.7×10^{-5}	7.0×10^{-6}
RO-5-4864	$>10^{-5}$	–	$>10^{-5}$	–

Values are apparent K_i, calculated as described under *Methods.* The following ligands had little or no effect on binding when tested at 10^{-4} M under the conditions employed ([^{3}H]FNZ concentration of 14–20 nM): Picrotoxin, bicuculline, glutamate, kainic acid, strychnine, amobarbital, chloropromazine, atropine, nicotine, d-tubocurarine, α-bungarotoxin, adenosine, 3-isobutyl-1-methyl-xanthine, theophylline, ε-aminocaproyl, concanavalin A (0.5 mg/ml), wheat-germ agglutinin (0.2 mg/ml), wax-bean agglutinin (0.1 mg/ml).

in the presence of GABA, in good agreement with the K_D calculated from binding isotherms (table IV). Kinetic measurements were not performed at higher temperatures since the reactions were too rapid to permit accurate determinations.

Ligand Interaction with the Benzodiazepine-Binding Sites

The pharmacological profile of calf cortex benzodiazepine-binding sites is presented in table V. Inhibition constants, K_i, of various ligands were estimated from the relation:

$$K_i = \frac{ED_{50}}{1 + \frac{[L]}{K_D}},$$

where ED_{50} is the concentration of ligand that displaces 50% of [^{3}H]FNZ under the assay conditions, K_D is the apparent dissociation constant of [^{3}H]FNZ determined from binding isotherms (see table IV), and [L] is the concentration of [^{3}H]FNZ. It appears that the pharmacological profile of [^{3}H]FNZ-binding sites correlates well with the pharmacological potency of various benzodiazepines as determined *in vivo* in mammals (17). Thus flunitrazepam was most potent whereas RO-5-4864, which is essentially inactive *in vivo,* did not displace [^{3}H] FNZ even at 10^{-5} M (table V). It is also seen that the relative potency of various benzodiazepines remained essentially unaltered whether tested in H or M preparations at 0°C or 37°C.

None of the drugs tested, aside from benzodiazepines, significantly inhibited [^{3}H] FNZ binding (table V). As described above, GABA, (±)-trans-3-aminocyclopentane carboxylic acid and to a lesser extent β-alanine enhanced binding.

Discussion

Properties of particulate benzodiazepine-binding sites from rat brain have recently been described (1, 2, 14, 15, 20). Calf brain offers an advantage as a starting material for biochemical investigations of these drug receptors because of the large amount of tissue available. A calf cortex homogenate contains about 1 pmol binding sites per mg protein, and a single cortex thus contains about 20 nmol binding sites. Assuming 100,000 daltons as molecular weight per site (a hypothetic value to demonstrate order of magnitude), one gets about 2 mg sites per single cortex.

Rat brain benzodiazepine receptors were studied at 0–4°C because specific binding of the ligand employed, [^{3}H] diazepam, was reported to be greatly reduced at physiological temperatures (2, 15). Binding of [^{3}H] FNZ to rat brain preparations at 0°C was also recently reported (22). The use of [^{3}H] FNZ enables one to study the benzodiazepine putative receptor at physiological temperatures. As seen in studies with [^{3}H] diazepam, the affinity of the receptor decreased with increasing temperature but even at 37°C the K_D for [^{3}H] FNZ was still about 10^{-8} M and non-specific binding at that concentration was very low. The change in affinity may be due either to the presence of an inhibitory endogenous ligand, which dissociates from the receptor at low temperature, or to a conformational change in the receptor itself. Extensive washings of membranes did not cancel the temperature-dependent alteration in affinity. It is therefore plausible to assume that the decrease in affinity observed at high temperatures is due to a conformational change in the receptor itself.

Washing the membranes did cause a consistent change in affinity as compared to the homogenate. A factor present in the supernatant increased affinity, and this modulation could be mimicked by GABAergic agonists and was blocked by bicuculline. The supernatant factor is probably GABA itself, since the concentration of this putative inhibitory transmitter in calf brain homogenate is 0.2–0.4 mM (10), more than enough to account for the observed effect *in vitro*. The presence of endogenous ligands that seem to inhibit benzodiazepine binding has recently been reported from several laboratories (7, 9, 13, 19). The observation that calf cortex high-speed supernatant enhances binding indicates that in whole-cortex homogenate the endogenous activator(s), e.g. GABA, is present at a concentration that overcomes the action of the endogenous inhibitors. This is very probably due to disruption of tissue compartmentization and post-mortem increase in GABA or other ligands (10). It is plausible to assume that the affinity of the benzodiazepine-binding sites *in vivo* is regulated by a sensitive interplay of endogenous inhibitors and activators, all of which may be part of the GABAergic system (3, 5, 8, 21, 22). The results also indicate that caution should be practised in interpreting results of receptor binding studies performed in non-washed particulate preparations, since such preparations may contain endogenous modulators (see for example ref. 7).

Modulation of benzodiazepine-binding sites by GABAergic ligands has recently been described in rat brain preparations at 0–4°C (21, 22), and in calf brain preparations at 37°C (5). It was reported that GABA increases the on-reaction of diazepam with its receptor but has no effect on the off-reaction (22). Our data indicate that the dissociation of [^{3}H]FNZ from the putative receptor is slowed down by GABA, an observation that is compatible with the finding that the dissociation of [^{3}H]FNZ from its receptor is slower in a homogenate than in a washed membrane preparation (data not shown).

Analysis of the binding data at all the temperatures tested revealed a single class of binding sites of [^{3}H]FNZ at the concentration range of 0.1–20 nM. However, this does not necessarily imply that there exists only a single class of benzodiazepine receptors in calf cortex. The high affinity of [^{3}H]FNZ may indicate that it is an antagonist (4). Heterogeneity of receptors may be masked by binding studies performed with potent antagonists (see, for example, ref. 6). Future studies with the yet unidentified agonists for the benzodiazepine receptors may thus reveal heterogeneity of these sites.

In the absence of detergents, essentially all benzodiazepine-binding sites appear to be associated with membranes. This association is apparently not solely electrostatic because high ionic strength did not seem to deplete the membranes of their activity. Various detergents released a substantial part of the binding activity into a high-speed supernatant. Further experiments are required to establish whether the binding observed in such preparations could be attributed to solubilized receptors or to aggregates or small membrane fragments. Detergent treatment may lead to further purification of benzodiazepine-binding sites, and thus to elucidation of their molecular structure and putative association with molecular constituents of the GABAergic system.

Summary

[^{3}H]Flunitrazepam ([^{3}H]FNZ) specifically binds to a single class of sites in calf cortex homogenate at a level of about 1 pmol/mg protein. Essentially all binding sites sediment after a 30 min centrifugation at 20,000 × g. The affinity of the sites decreases with increasing temperature both in a homogenate and in a washed membranes preparation. Binding sites in the washed membranes preparation display lower affinity than those in the homogenate (e.g., K_D of 14.8 nM vs. 7.6 nM respectively at 37°C), but membrane-sites affinity can be increased by aliquots of a high-speed supernatant as well as by GABA ($>10^{-7}$ M). GABA has only a small effect on the on-reaction but slows the off-reaction. The binding sites require chloride ions (~100 mM) for optimal activity, are inhibited by Hg^{+2}, and are partially released into a high-speed supernatant by several detergents.

References

1 Bosmann, H.B.; Case, K.R., and DiStefano, P.: Diazepam receptor characterization: specific binding of a benzodiazepine to macromolecules in various areas of rat brain. FEBS Lett. *82:* 368–372 (1977).

2 Braestrup, C. and Squires, R.F.: Specific benzodiazepine receptors in rat brain characterized by high-affinity [^{3}H] diazepam binding. Proc. natn. Acad. Sci. USA *74:* 3805–3809 (1977).

3 Costa, E.; Guidotti, A.; Mao, C., and Suria, A.: New concepts on the mechanism of action of benzodiazepines. Life Sci. *17:* 167–186 (1975).

4 Costa, T.; Rodbard, D., and Pert, C.B.: Is the benzodiazepine receptor coupled to a chloride anion channel? Nature, Lond. *277:* 315–317 (1979).

5 Dudai, Y.: Modulation of benzodiazepine binding sites in calf cortex by an endogenous factor and GABAergic ligands. Brain Res. *167:* 422–425 (1979).

6 Fisher, A.; Grunefeld, Y.; Weinstock, M.; Gitter, S., and Cohen, S.: A study of muscarinic receptor heterogeneity with weak antagonists. Eur. J. Pharmacol. *38:* 131–139 (1976).

7 Greenlee, D.V.; Van Ness, P.C., and Olsen, R.W.: Endogenous inhibitor of GABA binding in mammalian brain. Life Sci. *22:* 1653–1662 (1978).

8 Guidotti, A.; Toffano, G., and Costa, E.: An endogenous protein modulates the affinity of GABA and benzodiazepine receptors in rat brain. Nature, Lond. *275:* 553–555 (1978).

9 Karobath, M.; Sperk, G., and Schonbeck, G.: Evidence for an endogenous factor interfering with [^{3}H] diazepam binding to rat brain membranes. Eur. J. Pharmacol. *49:* 323–326 (1978).

10 Lovell, R.A.; Elliot, S.J., and Elliott, K.A.C.: The γ-aminobutyric acid and factor I content of brain. J. Neurochem. *10:* 479–488 (1963).

11 Lowry, O.H.; Rosebrough, N.H.; Farr, A.L., and Randall, R.J.: Protein measurement with the Folin phenol reagent. J. biol. Chem. *193:* 265–275 (1951).

12 Mackerer, C.R. and Kochman, R.L.: Effects of cations and anions on the binding of [^{3}H] diazepam to rat brain. Proc. Soc. exp. Biol. Med. *158:* 393–397 (1978).

13 Marangos, P.J.; Paul, S.M.; Greenlaw, P.; Goodwin, F.K., and Skolnick, P.: Demonstration of an endogenous, competitive inhibitor(s) of [^{3}H] diazepam binding in bovine brain. Life Sci. *22:* 1893–1900 (1978).

14 Möhler, H. and Okada, T.: Benzodiazepine receptor: Demonstration in the central nervous system. Science, N.Y. *198:* 849–851 (1977).

15 Möhler, H. and Okada, T.: Properties of [^{3}H] diazepam binding to benzodiazepine receptors in rat cerebral cortex. Life Sci. *20:* 2101–2110 (1977).

16 Nielsen, M.; Braestrup, C., and Squires, R.F.: Evidence for a late evolutionary appearance of brain-specific benzodiazepine receptors: an investigation of 18 vertebrate and 5 invertebrate species. Brain Res. *141:* 342–346 (1978).

17 Randall, L.O.; Schallek, W.; Sternbach, L.H., and Ning, R.Y.: Chemistry and pharmacology of 1,4-benzodiazepines; in Maxwell, Psychopharmacological agents; vol. 3, pp. 175–281 (Academic Press, New York 1974).
18 Segal, M.; Sims, K., and Smissman, E.: Characterization of an inhibitory receptor in rat hippocampus: a microiontophoretic study using conformationally restricted amino acid analogues. Br. J. Pharmac. *54:* 181–188 (1975).
19 Skolnick, P.; Marangos, P.J.; Goodwin, F.K.; Edwards, M., and Paul, S.: Identification of inosine and hypoxanthine as endogenous inhibitors of [^{3}H] diazepam binding in the central nervous system. Life Sci. *23:* 1473–1480 (1978).
20 Squires, R.F. and Braestrup, C.: Benzodiazepine receptors in rat brain. Nature, Lond. *266:* 732–734 (1977).
21 Tallman, J.F.; Thomas, J.W., and Gallager, D.W.: GABAergic modulation of benzodiazepine binding site sensitivity. Nature, Lond. *274:* 383–385 (1978).
22 Wastek, G.J.; Speth, R.C.; Reisine, T.D., and Yamamura, H.I.: The effect of γ-aminobutyric acid on [^{3}H] flunitrazepam binding in rat brain. Eur. J. Pharmacol. *50:* 445–447 (1978).
23 Zbinden, G. and Randall, L.: Pharmacology of benzodiazepines: Laboratory and clinical correlation; in Garattini and Shore, Advances in pharmacology; vol. 5, pp. 213–291 (Academic Press, New York 1967).

Dr. Y. Dudai, Department of Neurobiology, The Weizmann Institute of Science, Rehovot (Israel)

Prog. biochem. Pharmacol., vol. 16, pp. 109–118 (Karger, Basel 1980)

Does Neuroleptic Blocking of Dopamine Receptors Continue After Chronic Treatment?[1]

R.H. Belmaker, H. Dasberg and R.P. Ebstein
Jerusalem Mental Health Center–Ezrath Nashim, Jerusalem

Introduction

Recent work has suggested that the mode of action of neuroleptic drugs in schizophrenia involves blockade of dopamine neurotransmitter function (22). Animal evidence for this hypothesis includes neuroleptic inhibition of dopamine-stimulated adenylate cyclase (10, 11, 13), binding of neuroleptics to dopamine receptors in brain homogenates (19), and increased dopamine turnover after neuroleptic administration (6). Evidence that neuroleptics block dopamine transmission in humans includes the increased accumulation of a dopamine metabolite, homovanillic acid (HVA), in human cerebrospinal fluid (CSF) of patients under treatment with neuroleptics (9).

A difficulty with this theory is the finding of *Post and Goodwin* (17), who reported that after 3–10 weeks of neuroleptic treatment HVA accumulation in CSF returns to baseline while clinical improvement continues. A similar question regarding the relevance of dopamine blockade to the clinical efficacy of neuroleptic drugs in schizophrenia derives from studies of dopamine receptor supersensitivity. Since 3 weeks of neuroleptic treatment in animals leads to a marked increase in the behavioral response to dopamine agonists (14) and also in the number of dopamine receptor sites (5), it would seem that dopaminergic synapses are capable of compensating for neuroleptic receptor blockade. Such compensation must be reconciled with the lack of tolerance to the clinical efficacy of neuroleptics.

If the continuing clinical efficacy of neuroleptics is indeed related to dopamine receptor blockade one could hypothesize that: a) the return of HVA

[1] Supported by a grant from the Scottish Rite Schizophrenia Research Program, NMJ, USA.

The authors would like to thank *Dalia Pickholz* for technical assistance and *Keren Rose* for secretarial help.

turnover to baseline after 3–10 weeks of neuroleptic treatment is a function of presynaptic adaptation only, perhaps due to exhaustion of the presynaptic compensatory increase in dopamine release, and does not reflect any change in postsynaptic dopamine receptor blockade; and b) the increase in dopamine receptor number after 3 weeks of neuroleptic treatment is functionally ineffective in overcoming pharmacologic receptor blockade, and is reminiscent of the sprouting of axons in a severed human peripheral nerve, which relates to physiology important in some levels of phylogeny or in some periods of embryogenesis but which is only rarely of importance in the adult human being.

To test the above hypotheses investigators must study the effects of more than 3 weeks of neuroleptic therapy and must attempt to find possible postsynaptic indicators of dopamine blockade and to quantitate their significance. Cyclic AMP has been reported to be the product of stimulation of the dopamine receptor (13), and thus CSF cyclic AMP might be expected to be chronically reduced in schizophrenic patients under neuroleptic treatment if effective dopamine receptor blockade is a continuing phenomenon. In animals, the finding of increased dopamine receptor number after 3 weeks of neuroleptic therapy should be restudied after longer periods of neuroleptic therapy, to shed light on whether these receptor number changes limit themselves in such a way as to fail to functionally overcome neuroleptic blockade.

I The Effect of Neuroleptic Treatment on Human CSF Cyclic Nucleotides

A blood-brain barrier has been demonstrated for human cyclic AMP (4), and the low levels of CSF cyclic AMP in cerebral atrophy suggest a brain origin for this compound (23). *Sebens and Korf* (18) reported that intraventricular administration of dopamine or noradrenaline in rats raises CSF cyclic AMP, suggesting an origin in activity of neurotransmitter-sensitive adenylate cyclase. High-dose propranolol therapy of schizophrenic patients did not reduce CSF cyclic AMP, whereas patients with Parkinson's disease were found to have 40–50% reductions in CSF cyclic AMP (1). These findings indicate that brain dopamine-sensitive adenylate cyclase may make a significant contribution to human lumbar CSF cyclic AMP (1). Cyclic GMP in CSF has been much less studied, although in several tissues it may be the product of cholinergic receptor activation (16). It has not been found that CSF cyclic AMP and cyclic GMP levels can distinguish drug-free schizophrenic patients from normal controls (2), although a subgroup of core Kraepelinian schizophrenic patients may be characterized by high CSF cyclic AMP (24).

The effect of 4–12 weeks of neuroleptic treatment on CSF cyclic AMP and cyclic GMP was reported by *Biederman et al.* (3) and by *Ebstein et al.* (7) in 10

Table I. The effect of neuroleptic treatment in schizophrenia on spinal fluid cyclic AMP and cyclic GMP levels

Patient	Age	Sex	Diagnosis	Dosage	Weeks of treatment	Cyclic AMP before	Cyclic AMP after	Cyclic GMP before	Cyclic GMP after	Improvement rating
Responders										
1	26	F	A	600	11	30	20	1.2	2.5	+
2	26	M	R	600	4	25	27	1.2	4.2	+
3	59	M	R	200	10	20	11	1.0	3.5	+
4	43	M	R	200	6	19	11	3.0	3.7	+
5	23	M	A	1,200	14	37	36	1.5	3.2	++
6	63	F	R	200	8	34	18	3.0	2.7	++
7	16	M	A	840	5	38	22	3.2	2.5	+
8	23	M	R	1,000	12	48	39	1.2	2.5	+
Mean of responders only				605	8.7	31.4[1]	22.8[1]			
Nonresponders										
9	18	F	A	400	4	63	98	3.2	3.0	–
10	47	M	A	200	4	30	29	3.0	4.2	–
Mean overall	34.4		5A, 5R	544	7.8	34.4	31.1	2.03[2]	3.05[2]	

A = acute schizophrenia; R = recurrent schizophrenia. Dosage in chlorpromazine equivalents, mgm per day. Cyclic nucleotides in picomoles per cc of CSF. Improvement as none (–), moderate (+), or marked (++).

[1] $p < 0.05$. [2] $p < 0.01$.

schizophrenic patients who were studied drug-free on hospital admission and who agreed to repeat CSF sampling. Table I presents these results. There was a significant decline in CSF cyclic AMP only in those who responded clinically to neuroleptic treatment (8 patients), or in those who were treated for 5 weeks or more with neuroleptics (7 patients). The cyclic AMP decline is consistent with continued blockade of dopamine-sensitive adenylate cyclase after continued neuroleptic treatment, despite the reported return of CSF HVA levels to baseline. CSF cyclic GMP *increased* significantly in all 10 patients. The CSF cyclic GMP rise after neuroleptic treatment is consistent with concepts of restored acetylcholine-dopamine balance after such therapy (12) and with data suggesting that dopamine neurons form inhibitory synapses on cholinergic neurons in the striatum.

Smith et al. (21) were unable to find differences in CSF cyclic AMP and cyclic GMP levels between drug-free patients and patients taking pimozide. The difference between the results of *Smith et al.* (21) and those of *Biederman et al.* (3) and *Ebstein et al.* (7) may be due to several factors: a) *Smith et al.* (21) used pimozide which is a poorer blocker of dopamine-sensitive adenylate cyclase than might be expected from its clinical potency; b) *Smith et al.* (21) studied patients who had been treated with neuroleptics for widely varying periods of time (4–40 days), and it is conceivable that increases in dopamine turnover early in treatment might overcome receptor blockade and result in normal CSF cyclic AMP levels; c) *Smith et al.* (21) do not report their data in a before-after design. Assay-to-assay variability in binding and radioimmunoassays makes it imperative that all samples be assayed simultaneously and that individual patients be their own controls if possible; d) *Smith et al.* (21) did not separate neuroleptic non-responders from responders. *Bowers* (personal communication) has been able to confirm the reduction in CSF cyclic AMP with neuroleptic therapy.

II The Effect of Oral Haloperidol on Rat Striatal Dopamine Receptor Number and Binding Constant

Chronic neuroleptic treatment in rats has been shown to lead to a supersensitivity of the dopamine system as measured behaviorally (14) or as measured by increase in radioligand-assayed receptor number (5). Practically all experiments using biochemical measures have been carried out for a period of 3 weeks or less of neuroleptic therapy (5, 14, 15). Although often used as a model of tardive dyskinesia, haloperidol therapy for 3 weeks in rats is probably short-term even for investigating treatment effects. It should be remembered than an average psychotic episode in humans is treated with neuroleptic drugs for several months, and tardive dyskinesia typically requires years of neuroleptic treatment to develop (20). Thus the increased receptor number after 3 weeks of haloperi-

Table II. The effect of chronic haloperidol on the number of ^{3}H-spiroperidol binding sites in striatum

	Number of receptors (pmole/g tissue)		% change
	control	haloperidol	
3 weeks	32.27 ± 2.30 (± SEM, n = 5)	51.3 ± 3.95 (± SEM, n = 5)	+ 58.9[1]
10 weeks	26.51 ± 2.35 (± SEM, n = 6)	60.56 ± 7.87 (± SEM, n = 5)	+ 128[2]

[1] $t = 2.81$ Student's t test, $p < 0.05$.
[2] $t = 3.04$, $p < 0.01$.
Numbers in parentheses are the number of individual rat striatum assayed. Behavioral supersensitivity as measured by amphetamine (5 mg/kg)-induced stereotypy developed in a parallel group of haloperidol-fed animals ($t = 4.94$ Student's t, $p < 0.01$).

Table III. The effect of chronic haloperidol on the dissociation constant (K_D) of ^{3}H-spiroperidol to rat striatal membranes

	K_D (nM)		% change
	control	haloperidol	
3 weeks	0.41 ± 0.07 (± SEM, n = 5)	0.52 ± 0.05 (± SEM, n = 5)	+ 26.8 (n.s.)
10 weeks	0.55 ± 0.05 (± SEM, n = 6)	1.41 ± 0.24 (± SEM, n = 5)	+ 156[1]

[1] $t = 2.61$ Student's t test, $p < 0.05$.
Numbers in parentheses are number of individual rat striatum assayed.

dol treatment in rats may parallel changes that take place during continued clinically effective therapy with dopamine blocking drugs. If the increased dopamine receptor number is functionally not significant in the presence of neuroleptic receptor blockade, one must hypothesize that dopaminergic transmission will not have reachieved equilibrium and that further but limited increases in dopamine receptor number might be expected with further neuroleptic treatment.

The effect of 3 and 10 weeks of haloperidol treatment on striatal dopamine receptor number and affinity constant are presented in tables II and III (8). Table II shows that the increase in receptor number reported after 3 weeks not only continues after 10 weeks but becomes more marked. Table III shows the effect of haloperidol treatment on the affinity constant. No changes in binding constant were reported by others after 3 weeks (5, 14, 15), and were not found in the present study. However, after 10 weeks of haloperidol treatment a significant increase in K_D was seen, implying a significant decrease in affinity of the receptor for spiroperidol. While further studies are clearly indicated to determine whether this conformational change is specific to the antagonist state of the receptor, such specificity seems unlikely. More probably, the decreased affinity for ^{3}H-spiroperidol represents the appearance of partially defective or 'immature' receptors as the number of receptors increases. This would be analogous to the appearance of immature leukocytes during infection-induced leukocytosis. The fact that such poor-affinity receptors appear only after more than 3 weeks of therapy and after the significant increase in receptor number suggests that a reserve exists of fully effective receptors, and that this reserve can be exhausted with further increase in receptor number. Continued increase in receptor number with 10 weeks of haloperidol treatment as compared with 3 weeks at the same dose suggests that equilibrium or basal levels of dopamine neurotransmission were not achieved by 3 weeks of neuroleptic treatment; the appearance of a decrease in receptor affinity for ^{3}H-spiroperidol at 10 weeks suggests that receptor mechanisms may not be capable of truly compensating for neuroleptic-induced blockade. Such data is consistent with the chronically reduced levels of CSF cyclic AMP found in patients responding to neuroleptic therapy.

Figure 1 shows differences in dopamine receptor number of individual rats in response to 3 and 10 weeks of haloperidol treatment. The variance at 10 weeks is significantly greater in the haloperidol-treated rats than in the controls, and it is clear that some rats have a more than 400% increase in dopamine receptor number. It is tempting to speculate that such individual differences exist in humans as well, and that a brisk response of this kind in some patients *could* overcome dopamine blockade and result both in lack of clinical improvement and in an absence of a decline in CSF cyclic AMP in neuroleptic non-responders. Figures 2 and 3 illustrate individual differences in response of CSF cyclic AMP and cyclic GMP to neuroleptic treatment in humans (as in table I). The variability of this phenomenon, too, emphasizes the need for correlation with clinical response.

Further research in this area should emphasize long-term treatment and clinical correlation, i.e. biological heterogeneity among individuals of the same species. The human study of CSF cyclic AMP described above clearly has too few patients and requires replication. However, replication should pay careful attention to possible issues of time course and clinical correlates. Studies early in neuroleptic treatment may find CSF cyclic AMP unchanged because presynaptic increases in dopamine release overcome receptor blockade; alternatively, studies of chronic schizophrenic neuroleptic nonresponders may select a group with very active receptor adaptational mechanisms and thus an absence of CSF cyclic AMP decline with neuroleptics. Animal studies of neuroleptic treatment should also be carried out over a longer period, especially since administration of haloperidol in food reduces an otherwise heavy burden of daily injection. Pharmacokinetic questions need to be considered, such as whether individual animals

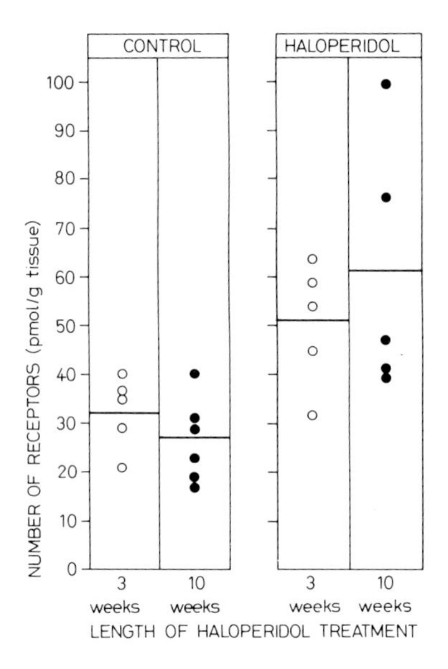

Fig. 1. Distribution of receptor number after chronic haloperidol treatment in rats. Rat food containing 0.01% haloperidol was prepared by grinding regular rat pellets to a fine powder and thoroughly mixing with drug. Control rats received the same powdered food without haloperidol. Male Sabra strain rats were used in all experiments and weight gain on this diet was normal. Approximate daily oral dose was 3 mg of haloperidol per rat. Rats were sacrificed 4 days after cessation of haloperidol feeding and the striatum was dissected and stored at –70°C until assayed. The binding of ^{3}H-spiroperidol to striatal homogenate was determined as described by *Burt, Creese and Snyder* (5). Striatum was homogenized using a glass-teflon homogenizer in 100 volumes of 50 nM Tris buffer pH 7.7 containing the following components: 120 mM NaCl, 5 mM KCl, 2 mM $CaCl_2$, 1 mM $MgCl_2$, 0.1% ascorbic acid and 10 μM pargyline. The membranes were collected by centrifugation (50,000 × g for 10 min) and resuspended in 285 volumes (original wet weight) of buffer. The reaction mixture contained 800 μl membrane suspension, 100 μl ^{3}H-spiroperidol (NEN, 23 Ci/mM) from 0.1 to 1.0 nM (5 different concentrations) and either 100 μl 0.1% ascorbic acid or 10 μM dopamine (blank) in 0.1% ascorbic acid. After 10 min incubation at 37°C the reaction was stopped by rapidly filtering through Whatman GF/B glass fiber filters and washing with ice-cold buffer (2 × 10 ml). The filters were counted in 10 cc Instagel after shaking vigorously for 2 h. Specific binding of ^{3}H-spiroperidol was calculated as the number of counts in excess over the dopamine blank. The number of receptor sites and the K_D was determined by Scatchard plot for each individual rat striatum. Plots were fitted with a standard computer program. There is a significantly increased variance ($F = 9.31$, $p < 0.05$) in the number of receptor binding sites after 10 weeks of haloperidol treatment compared with control-treated animals. After only 3 weeks haloperidol there is no significant difference between the variance of the drug and control-treated animals ($F = 2.94$, n.s.). In striatum ^{3}H-spiroperidol binding is a good measure of specific dopamine binding sites.

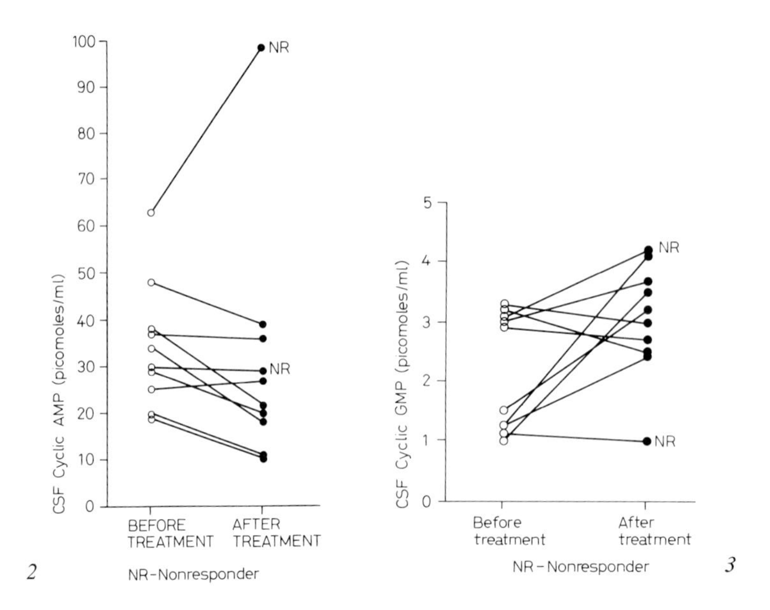

Fig. 2. Variability of the effect of neuroleptic treatment on CSF cyclic AMP.
Fig. 3. Variabilty of the effect of neuroleptic treatment on CSF cyclic GMP.

receive different dosages by oral feeding. Apparent decreases in affinity constant could theoretically result from continued presence in the preparation of the blocking drug, although this seems unlikely after the dilution and washing inherent in the binding assay, and in any case such effects should not differ between 3 and 10 weeks of haloperidol treatment.

Summary

Neuroleptic treatment of schizophrenia reduces CSF cyclic AMP in clinical responders after a mean of 8 weeks of treatment. This is consistent with continued effective dopamine receptor blockade by these drugs. Rat striatal ^{3}H-spiroperidol binding sites increase more after 10 weeks of haloperidol therapy than after 3 weeks, suggesting that physiological equilibrium has not yet been reached by 3 weeks. The rat striatal receptor affinity for ^{3}H-spiroperidol is decreased after 10 weeks but not after 3 weeks. The decreased affinity represents a new phenomenon that may be important in explaining long-term continuing neuroleptic effectiveness despite the increased receptor number. The decreased affinity of rat dopamine binding sites for ^{3}H-spiroperidol after 10 weeks of neuroleptic treatment is consistent with the reduced human CSF cyclic AMP after a similar period of neuroleptic treatment.

References

1 Belmaker, R.H.; Ebstein, R.P.; Biederman, J.; Stern, R.; Berman, M., and van Praag, H.M.: The effect of L-dopa and propranolol on human CSF cyclic nucleotides. Psychopharmacology *58:* 307–310 (1978).

2 Biederman, J.; Rimon, R.; Ebstein, R.P.; Belmaker, R.H., and Davidson, J.T.: Cyclic AMP inthe CSF of patients with schizophrenia. Br. J. Psychiat. *130:* 64–67 (1977).

3 Biederman, J.; Rimon, R.; Ebstein, R.P.; Zohar, J., and Belmaker, R.H.: Neuroleptics reduce spinal fluid cyclic AMP in schizophrenic patients. Neuropsychobiology *2:* 324–327 (1976).

4 Brooks, B.R.; Engel, W.K., and Sode, J.: Blood-to-cerebrospinal fluid barrier for cyclic adenosine monophosphate in man. Archs Neurol., Chicago *34:* 468–469 (1977).

5 Burt, D.R.; Creese, I., and Snyder, S.H.: Antischizophrenic drugs: Chronic treatment elevates dopamine receptor binding in brain. Science, N.Y. *196:* 326–327 (1977).

6 Carlsson, A. and Lindquist, M.: Effect of chlorpromazine and haloperidol on formation of 3-methoxytryptamine and norepinephrine in mouse brain. Acta pharmac. tox. *20:* 140–144 (1963).

7 Ebstein, R.P.; Biederman, J.; Rimon, R.; Zohar, J., and Belmaker, R.H.: Cyclic GMP in the CSF of patients with schizophrenia before and after neuroleptic treatment. Psychopharmacology *51:* 71–74 (1976).

8 Ebstein, R.P.; Pickholz, D., and Belmaker, R.H.: Dopamine receptor changes after long-term haloperidol treatment in rats. J. Pharm. Pharmac. *31:* 558–559 (1979).

9 Fyro, B.; Wode-Helgodt, B.; Borg, S., and Sedvall, G.: The effect of chlorpromazine on homovanillic acid levels in cerebrospinal fluid of schizophrenic patients. Psychopharmacologia *35:* 287–294 (1974).

10 Greengard, P.: Biochemical characterization of the dopamine receptor in the mammalian caudate nucleus. J. psychiat. Res. *11:* 87–90 (1974).

11 Iversen, L.L.: Dopamine receptors in the brain. Science, N.Y. *188:* 1084–1089 (1975).

12 Janowsky, D.S.; El-Yousef, M.K., and Sekerke, H.J.: Antagonistic effects of physostigmine and methylphenidate in man. Am. J. Psychiat. *130:* 1370–1376 (1973).

13 Karobath, M. and Leitich, H.: Antipsychotic drugs and dopamine-stimulated adenylate cyclase prepared from corpus striatum of rat brain. Proc. natn. Acad. Sci. USA *71:* 2915–2918 (1974).

14 Klawans, H.L.; Hitri, A.; Nausieda, P.A., and Weiner, W.J.: Animal models of dyskinesia; in Hanin and Usdin, Animal models in psychiatry and neurology; pp. 351–363 (Pergamon Press, Oxford 1977).

15 Muller, P. and Seeman, P.: Brain neurotransmitter receptors after long-term haloperidol: Dopamine, acetylcholine, serotonin, α-noradrenergic and naloxone receptors. Life Sci. *21:* 1751–1758 (1977).

16 Palmer, G.C. and Duszynski, C.R.: Regional cyclic GMP content in incubated tissue of rat brain. Eur. J. Pharmacol. *32:* 375–379 (1975).

17 Post, R.M. and Goodwin, F.K.: Time-dependent effects of phenothiazines on dopamine turnover in psychiatric patients. Science, N.Y. *190:* 488–489 (1975).

18 Sebens, J.B. and Korf, J.: Cyclic AMP in cerebrospinal fluid: Accumulation following probenecid and biogenic amines. Expl Neurol. *46:* 336–344 (1975).

19 Seeman, P.; Chau-Wong, M.; Tedesco, J., and Wong, K.: Brain receptors for antipsychotic drugs and dopamine. Direct binding assays. Proc. natn. Acad. Sci. USA *72:* 4376–4380 (1975).

20 Shader, R.I. and Dimascio, A.: Psychotropic drug side effects (Williams & Wilkins, Baltimore 1970).

21 Smith, C.C.; Tallman, J.F.; Post, R.M.; van Kammen, D.P.; Jimmerson, D.C., and Brown, G.L.: An examination of baseline and drug-induced levels of cyclic nucleotides in the cerebrospinal fluid of control and psychiatric patients. Life Sci. *19:* 131–136 (1976).
22 Snyder, S.H.; Banerjee, S.P.; Yamamura, H.I., and Greenberg, D.: Drugs, neurotransmitters and schizophrenia. Science, N.Y. *184:* 1243–1253 (1974).
23 Vapaatalo, H.; Myllyla, V.; Heikkinen, E., and Hokkanen, E.: Cyclic AMP in CSF of patients with neurologic disease. New Engl. J. Med. *296:* 691–692 (1976).
24 Zohar, J.; Rimon, R.; Biederman, J.; Ebstein, R.P., and Belmaker, R.H.: Clinical correlates of CSF cyclic nucleotides in schizophrenia. Am. J. Psychiat. *135:* 253–255 (1978).

Dr. R.H. Belmaker, Jerusalem Mental Health Center–Ezrath Nashim, P.O.B. 140, Jerusalem (Israel)

Prog. biochem. Pharmacol., vol. 16, pp. 119–132 (Karger, Basel 1980)

Serotonin Receptor Site in Human Platelets from Control and Chlorpromazine Treated Subjects

B. Oppenheim, A. Hefez[1] *and M.B.H. Youdim*

Department of Pharmacology, Faculty of Medicine, Technion–Israel Institute of Technology, Haifa and [1]Department of Psychiatry, Rambam Medical Center, Haifa

Introduction

Human platelets show functional resemblance to isolated brain synaptosomes. They have certain clear advantages because they are intact human cells present in their physiological milieu, not detached from their natural connections as are synaptosomes. Moreover, the platelets are activated by neurotransmitters and the activation process can be carefully analyzed ultrastructurally and biochemically, under various physiological and non-physiological conditions.

Among its subcellular organelles, platelets possess osmiophilic vesicles in which serotonin (5-HT) is stored in association with non-metabolic ATP, calcium, magnesium and pyrophosphates (13, 22).

Unlike the nerve endings in the central nervous system (CNS), platelets do not synthesize their own 5-HT. The monoamine is accumulated in the storage vesicles by an active uptake process; anti-depressant agents that block monoamine uptake into brain synaptosomes (i.e. imipramine) also inhibit 5-HT uptake in human platelets (3).

Mitochondrial monoamine oxidase (MAO) in human platelets is of type B (29), similar to that found in human brains with regard to substrate specificities and inhibitor sensitivity (8, 16, 23).

Agents that deplete monoamine vesicle content in the brain also deplete platelet vesicular 5-HT (20). In addition, neuroleptics which inhibit presynaptic dopamine (DA) release from dopaminergic neurons in animal CNS (25, 26) and block DA-sensitive adenylate cyclase on the receptor site (6), affect platelet aggregation response (PAR) to 5-HT *in vitro* and *in vivo* in two distinct, almost opposite fashions (see below).

The present investigation was undertaken to study the mechanism of action of chlorpromazine (CPZ) as reflected by platelet function and to investigate the clinical significance of the CPZ-modified PAR.

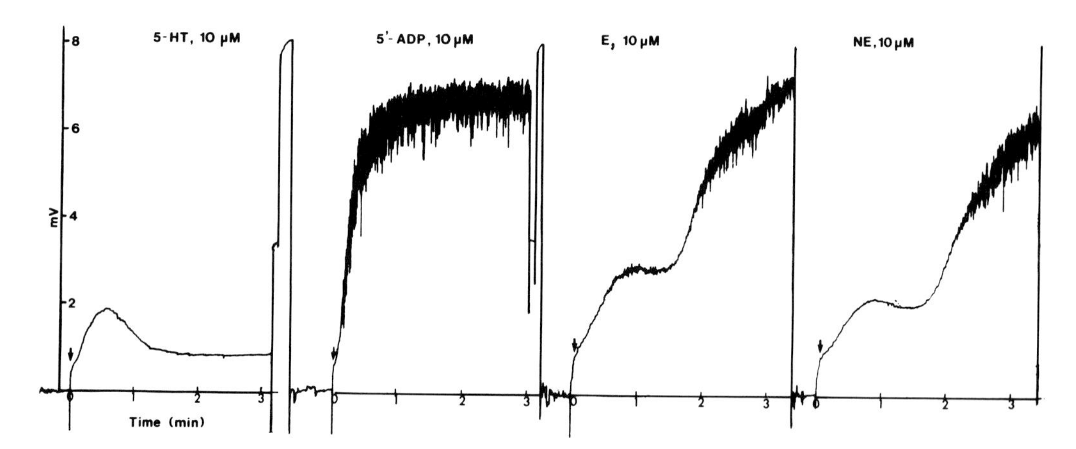

Fig. 1. PAR to monoamines and ADP. Human blood platelets from a control subject were isolated from whole venous blood by centrifugation at 200 × g at room temperature in a final concentration of sodium citrate of 12.9 mM. They were immediately transferred into plastic tubes and kept in a thermostatic bath at 37°C until their aggregation responses to the various inducers were determined. A supernatant, platelet-poor plasma was prepared by 10 min centrifugation at 2,000 × g and used as a reference in the photometric determinations. PAR to 5-HT, ADP, E and NE are illustrated. The curves indicate time-dependent changes in optical transmission. The aggregation inducers were added as indicated by the arrows.

Effect of Tryptolines on Platelet Aggregation Responses

In the experimental system, the process of human platelet aggregation was followed photometrically, as a reduction in the initial optical density of the platelet suspension. Rates of aggregation were dependent on concentration of inducers. The maximal responses for 5-HT, epinephrine (E), norepinephrine (NE) and ADP at saturating concentrations are shown in figure 1. The response to 5-HT, even at saturation, remained transient in 90% of human population, and the cells disaggregated without being able to aggregate again (1). Maximal PAR to E, NE and ADP were irreversible.

Since the initial observation of the transient 5-HT PAR by *Mitchell and Sharp* (15) and *Baumgartner and Born* (1), structural analogues of 5-HT and its antagonists have been tested. All were found to inhibit 5-HT PAR, the most potent inhibitor being the antagonist methysergide (3). None of the compounds tested (3) was able to induce a response similar to that of 5-HT.

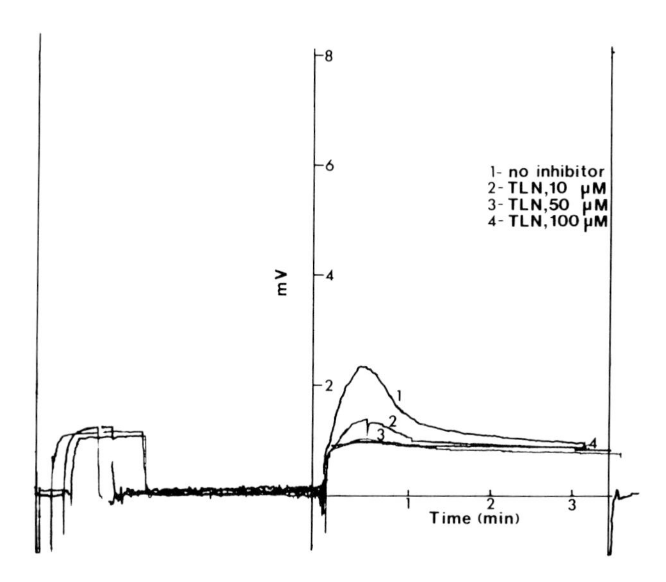

Fig. 2. Effect of tryptoline on PAR to 5-HT. Aliquots of 450 μl of platelet suspensions (prepared as described in fig. 1) were each treated for 3 min, with or without a given concentration of TLN, and the subsequent 5-HT PAR determined. The experimental curves are superimposed.

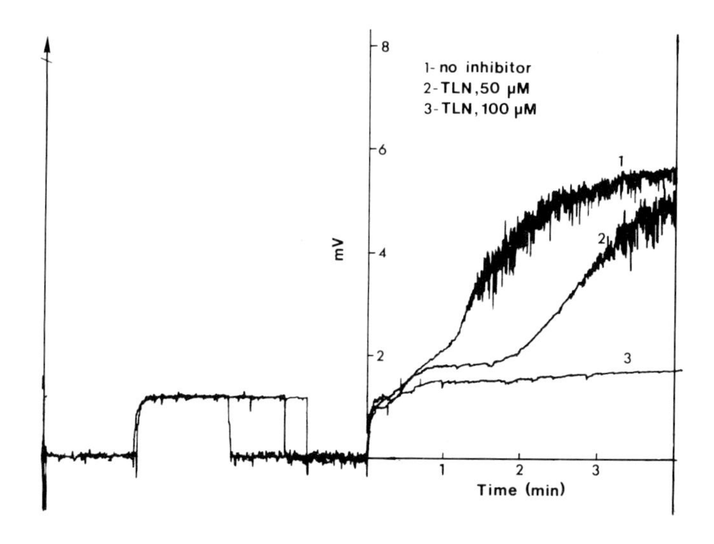

Fig. 3. Effect of tryptoline on PAR to epinephrine. Experimental details as in fig. 1.

Table I. Inhibition constants of tryptolines for human platelet aggregation responses to serotonin and epinephrine

Monoamine	Inhibitor	K_M (μM)	K_I (μM)
5-Hydroxytryptamine		0.36	
	5-hydroxytryptoline		3.3
	5-hydroxymethtryptoline		2.0
	tryptoline		9.6
Epinephrine		0.8	
	tryptoline		9.5

Each value is a mean of two to four determinations.
The range of concentrations of 5-HT and E used were 0.1–20 μM each. The effects of TLNs on PARs to 5-HT and E were tested after 3 min preincubation with varying concentrations of TLNs (1–100 μM). Determination of the kinetic constants was based on the values of initial rates of aggregation responses.

Tryptolines (tetrahydro-β-carbolines) (TLNs) are structural analogues of 5-HT. They are of great interest, because when administered in combination with an irreversible MAO inhibitor, they cause a syndrome of stereotyped hyperactivity in rats (12). This syndrome resembles that obtained when L-tryptophan is administered and followed by an irreversible MAO inhibitor, implying 5-HT agonist properties of TLNs. Moreover, it is possible that endogenous TLNs are synthesized *in vivo* in the brain from 5-HT (27). When tested as possible aggregation inducers, TLNs (1–200 μM) were found to be totally inert. However, they strongly inhibited PAR to the subsequent addition of 5-HT or E (figs. 2 and 3) and the inhibition constants, as presented in table I, were low. The 5-hydroxy derivatives exhibited the highest affinity for the system, as one would expect from compounds similar in structure to 5-HT (19). Besides being inhibitors of PAR to 5-HT and E, TLNs are also reversible inhibitors of 'type A' MAO activity in human platelets (30).

Another compound with 5-HT agonist properties is the 'methylated 5-HT' or 5-methoxy-N, N dimethyltryptamine (5-MNDTA). This compound also produced hyperactivity in experimental animals, which was enhanced if animals were pretreated with an irreversible MAO inhibitor. Like TLNs, this analogue at concentrations of up to 100 μM did not induce PAR but instead prevented expression of 5-HT PAR if preincubated with the platelets. These observations with structural analogues of 5-HT and its antagonists point to 5-HT as the sole inducer of transient PAR, and raise questions about the nature of the platelet 5-HT receptor.

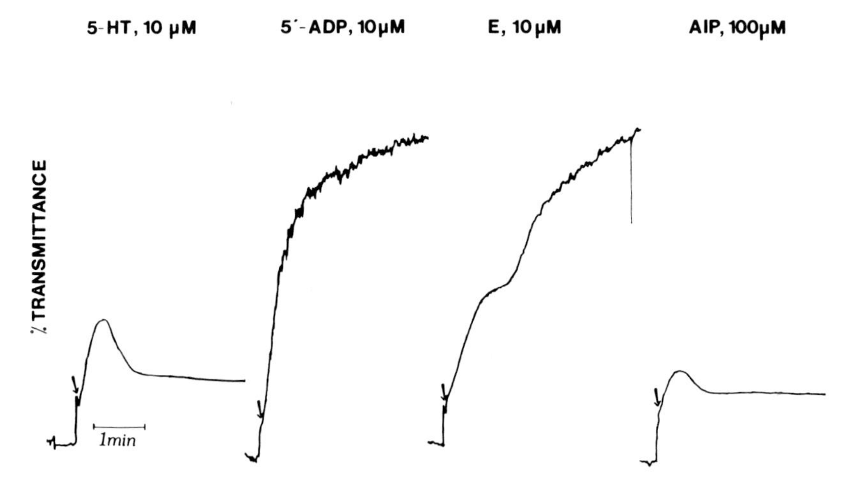

Fig. 4. PAR to 5-HT, ADP, E and AIP. A transient aggregation response of human control platelets to subsaturating concentrations of AIP is shown, together with the transient maximal response to 5-HT and the irreversible PAR to ADP and E.

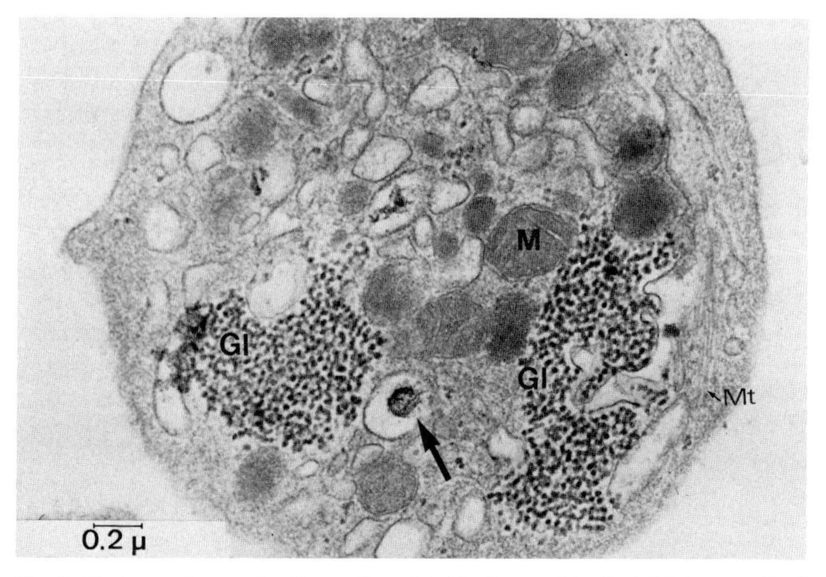

Fig. 5. Electron micrograph of a resting platelet. Human blood platelets were fixed with glutaraldehyde in 1.25 mM $CaCl_2$ and 50 mM Hepes buffer, according to a slightly modified procedure described by *Skaer* (28), post-fixed and stained using standard procedures. Symbols indicate the following: M – Mitochondrion, L – lipid, Gl – glycogen, Mt – microtubules. Arrows point to the 5-HT vesicles.

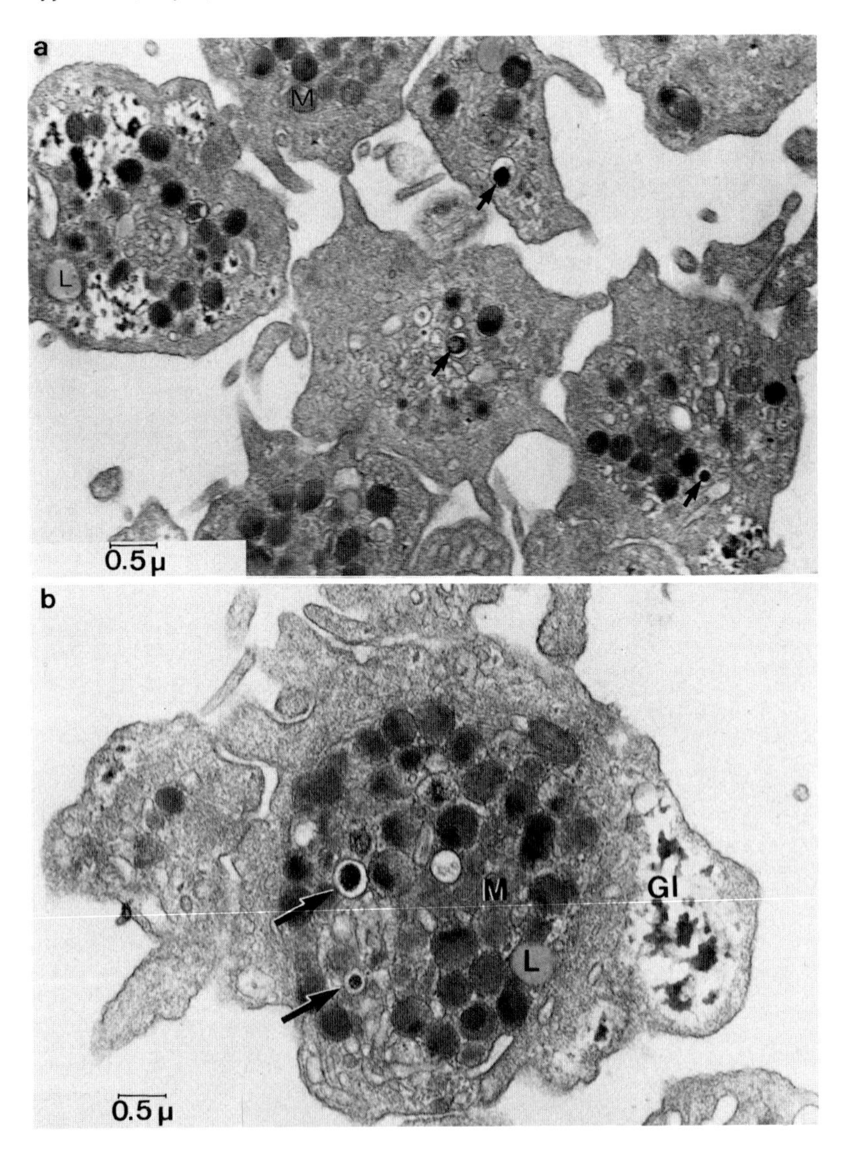

Fig. 6a. Electron micrograph of a group of platelets activated by 100 µM AIP for 30 sec. Experimental conditions and symbols as in fig. 5.

Fig. 6b. Electron micrograph of a human platelet treated with AIP. Notice glycogen degranulation and 5-HT-containing vesicle in the center. AIP concentration, 100 µM; time of incubation, 30 sec. Symbols as in fig. 5.

Effects of Adenylylimidodiphosphate on the Platelet System

While studying the action of a number of nucleotides on 5-HT PAR, a transient '5-HT-like' PAR was obtained by a preparation of adenylylimidodiphosphate (AIP) in a concentration range of 0.05–2 mM (17). As with 5-HT, PAR to AIP can be induced only once. The nitrogen between the β and γ position of the ATP analogue protects it from cleavage by phosphatases and makes this compound an ATPase inhibitor of moderate potency (K_1 = 0.8 mM, see, for example, ref. 11). The K_M for the transient PAR to AIP is 0.2–0.5 mM. A submaximal response is presented in figure 4. Drugs that inhibit PAR to 5-HT also inhibit the transient PAR to AIP (17).

Electron microscopic studies of platelets treated with AIP show an activation process similar to that induced by 5-HT, involving change in shape (into a 'spiny sphere'), disappearance of peripheral microtubules, centralization of vesicles and degranulation of glycogen (figs. 5, 6). Dense vesicles were not depleted after 30–60 sec of treatment with either 5-HT or AIP.

The transient nature of the monophasic aggregation response induced by AIP is explained by an additional effect of AIP, viz., the nucleotide was found to inhibit the second phase of aggregation responses to E, NE, ADP and 5-HT when a biphasic response was induced (see below). These and other observations (17, 18) indicate that AIP inhibits release of dense vesicle content (release I) by exocytosis, a possibility at present being examined.

AIP has been used in several systems as a tool to distinguish between the process of nucleotide binding and nucleotide dephosphoration. It is known that AIP is 97% pure, the major contaminant being adenylphosphoimidole. In platelet aggregation studies one cannot ignore the possibility of contaminants which can cause aggregation. For instance, the presence of 1% ADP in an ATP or AIP preparation can account for the transient '5-HT-like' PAR we have observed with AIP preparations. This problem is now being investigated by purifying all AIP preparations.

Interaction between ATP and Epinephrine

Even more remarkable was the effect of a combination of E and ATP in inducing PAR, at concentrations at which neither by itself caused measurable aggregation. Results so far obtained by applying both E and ATP (ATP added 40 sec or 3 min after E, as in figure 7) show a transient '5-HT-like' aggregation (fig. 7).

The synergism between E and adenine nucleotides in producing PAR is briefly dealt with elsewhere (18), but the most important question in this context is whether these transient '5-HT-like' PARs are mediated via a 5-HT receptor or by a separate site.

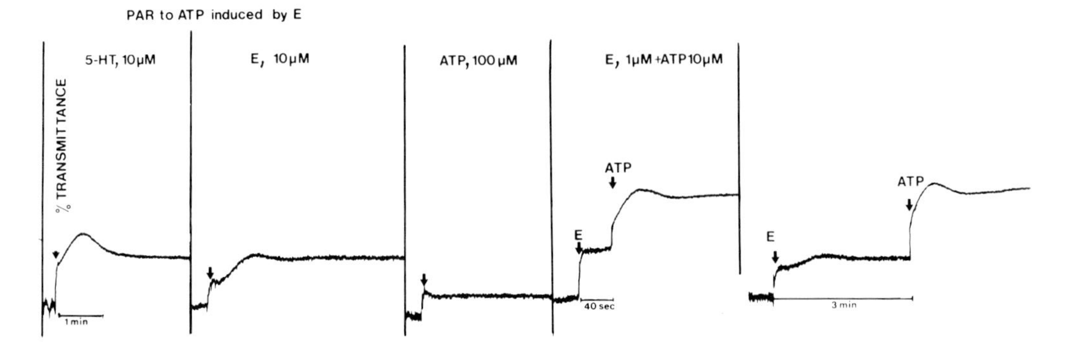

Fig. 7. PAR to ATP induced by epinephrine. Aggregation responses of platelets from a human subject with a monophasic E response (10 μM) and with no response to ATP (100 μM) are shown. Preincubation of platelets with E for 40 sec or 3 min caused a transient PAR upon subsequent addition of 10 μM ATP.

The in vitro *and* in vivo *Action of Chlorpromazine*

It is well known that chlorpromazine (CPZ) and related phenothiazines inhibit 5-HT PAR *in vitro* (4, 14). Indeed, as reported by *Mills and Roberts* (14) and later by *Boullin et al.* (4), this is the immediate *in vitro* effect of CPZ and was initially thought to be a useful measure of the drug efficacy in CPZ-treated patients. There is an urgent need for such an assay, since plasma levels of CPZ and its metabolites cannot be correlated with their clinical efficacy. However, the results *in vivo* are different: platelets from CPZ-treated patients did not have a diminished or inhibited PAR to 5-HT as one might expect from the presence of the drug in the plasma and from *in vitro* studies; on the contrary, in some patients the platelets exhibited an enhanced 5-HT PAR. This observation was originally published by *Boullin et al.* (5).

The appearance of irreversible 5-HT PAR is concentration-dependent: at saturation it is irreversible (fig. 8a), but reducing 5-HT to the range of K_M of the reversible PAR yields a transient 5-HT PAR (fig. 8a). This enhanced 5-HT PAR is sensitive to the same inhibitors that affect the transient 5-HT PAR, as if it were a response of a pre-existing unmasked or over-exposed site. The effect of TLN, for example, on irreversible PAR to 5-HT is shown in figure 8b. (Unlike the TLNs, AIP inhibits only the second phase of enhanced 5-HT aggregation.)

It was important to test the observation of enhanced 5-HT response in a clinical trial, to study the development of the enhanced PAR to 5-HT in CPZ-treated patients, and to re-examine this property as a biological assay for determining the efficacy of CPZ (10).

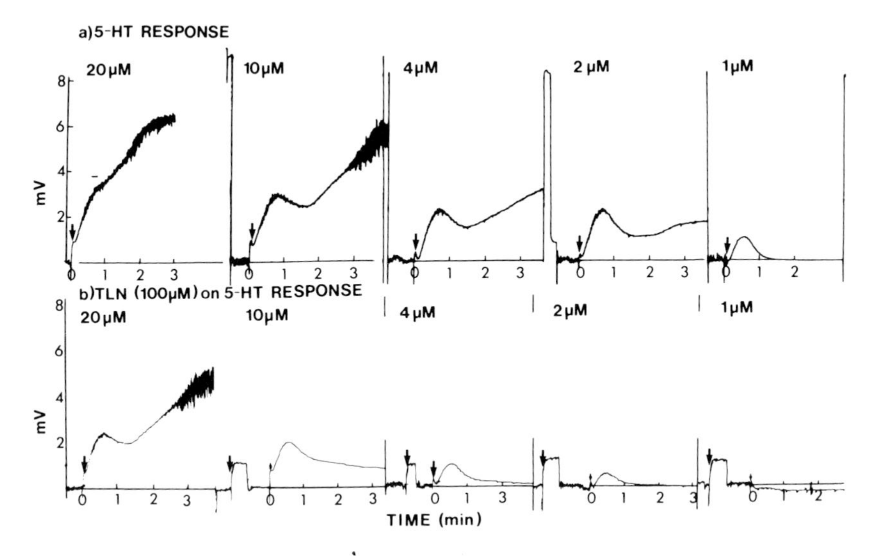

Fig. 8. Irreversible 5-HT PAR of a CPZ-treated patient and its inhibition by TLN: a) Different concentrations of 5-HT (1–20 μM) were used to induce PAR. A concentration-dependent increase in PAR to 5-HT from a reversible to an irreversible one is observed. b) The effect of 100 μM TLN on PARs, induced by various concentrations of 5-HT, is shown.

The first double-blind study included 22 patients (aged 19–40) having their first psychotic episode. In addition to CPZ, which was the sole neuroleptic, the patients were treated with electroconvulsive shock treatment (ECT) and with trihexyphenidyl to counteract extra-pyramidal side effects of CPZ. They were followed clinically for 3 months or until discharge from the hospital.

Weekly determination of platelet function and independent clinical assessment according to the Hopkins Psychiatric Rating Scale (HPRS) revealed the following: elevent patients were clearly defined as schizophrenic, according to the criteria of *Bleuler* (2), *Schneider* (24) and *Feighner et al.* (7). The changes in severity of psychoticism, conceptual disorder, paranoid ideation and global score were determined. Seven points of severity were attributed to each parameter and a decrease of 2 points was considered a clinical improvement. Initially, all patients had a 'reversible' 5-HT PAR. However, within 2–3 weeks, two groups of patients could be identified: in 6 patients, an enhanced biphasic 5-HT PAR developed, and this group included all patients assessed clinically as CPZ responders. In the remaining 5 patients, who were diagnosed as slow, poor or non-responders to CPZ, the reversible 5-HT PAR remained unchanged during 10–12 weeks of CPZ treatment (results not shown). Although only a few cases were tested, platelets with irreversible 5-HT PAR also showed considerably pronounced AIP responses. These observations are now being examined in a much larger group of patients in double-blind studies (10).

Thus, it may be possible to use platelet function as a bioassay: a) to distinguish at an early stage between CPZ responders and non-responders, so as to provide the non-responders with an alternative, more appropriate therapy; and b) to differentiate between subgroups in schizophrenia and to establish the efficacy of CPZ function in the central nervous system. However, further studies with other neuroleptics are required.

Discussion

Platelets display certain properties, such as uptake and release of 5-HT, which are very similar to those of synaptosomes (21). These functions are thought to be presynaptic in nature. It is not known whether the receptors responsible for platelet aggregation are of pre- or postsynaptic type. One may speculate on the basis of the present studies that these receptors are also of presynaptic nature: firstly, 5-HT agonists (TLNs and 5-MNDTA) antagonize the 5-HT aggregation response rather than mimic it, as one would expect from a postsynaptically acting agonist. Secondly, there is a close relationship with the uptake system, a presynaptic component, and uptake inhibitors such as imipramine and fluvoxamine (3, 19) are all 5-HT aggregation inhibitors.

There has been a difficulty in dissociating the 5-HT uptake system from that of the 5-HT receptor responsible for the transient aggregation (3). However, the availability of the compound AIP (structurally unrelated to 5-HT), which may mimic many properties of 5-HT with regard to aggregation, and which is not taken up by the platelets, may help to elucidate the relationship between the uptake system of 5-HT and the site responsible for the transient aggregation. However, further studies are in progress to determine whether the aggregation property of AIP is due to AIP or contaminants.

Platelet aggregation is not a biochemically defined reaction product, but a cell population behavior, and can be produced in a transient manner by 5-HT and AIP as well as by the combined effect of a low unsaturated (non-aggregating) concentration of E and ATP. The self-restricting 5-HT response can be enhanced into an irreversible 'true' aggregation response in a majority of schizophrenic patients responding to CPZ. If a similar enhancement should occur in the '5-HT-like' responses to AIP or E + ATP, we may have introduced a new approach to understanding the supersensitivity of receptor sites, due to CPZ treatment. Preliminary studies have indicated that this may be true for the AIP response. The possibility of a purinoreceptor involvement in the adenine nucleotide responses was ruled out.

Chlorpromazine, a molecule of many known metabolites and certainly having more than one pharmacological action, is known to have a dual effect in the brain: a) it inhibits presynaptic DA release from storage vesicles in the animal brain (25, 26); and b) it blocks postsynaptic DA receptor sites inhibiting DA-stimulated adenylate cyclase activity (6). In the platelet system, CPZ also has two effects: the immediate *in vitro* inhibitory effect on platelet aggregation – which is similar to the inhibition of central DA release – and its delayed accumulative hyperactivation (supersensitivity) of platelets by 5-HT (fig. 8a). There is an inevitable comparison with the effect of CPZ on the behavior of rats. A single dose of CPZ inhibited the hyperactive syndrome induced in rats by an irreversible MAO inhibitor and L-tryptophan or L-dopa (9a), whereas prolonged treatment of rats with CPZ resulted in a supersensitivity of serotonergic and dopaminergic neurons, as demonstrated in increased animal hyperactivity when compared with appropriate controls (9b).

Whether the supersensitivity in platelet aggregation observed in CPZ-treated patients reflects an increased number of '5-HT receptor' sites or a conformationally modified form of the same receptor, or even an extra-platelet component, is not known. On the basis of the results reported in the literature, in animal as well as human studies, one would assume that long-term treatment with CPZ unmasks a fraction of inaccessible 5-HT receptor sites.

For the purpose of functional analogy between blood platelets and brain synaptosomes, the short life span of platelets (8–9 days) initially seemed a serious drawback. However, if the platelet function, namely aggregation, demonstrates weekly changes reflecting changes produced centrally by drug treatment, it may prove to be an advantage and a useful tool in determining drug efficacy in patients treated with neuroleptics.

Summary

Human platelets are activated by epinephrine (E), norepinephrine (NE), ADP, and serotonin (5-HT) by a process that can be measured as cellular aggregation and at the completion of which 5-HT is released from storage vesicles. Among the three neurotransmitters, only the 5-HT response at its maximum concentration is transient (reversible), and in a given platelet preparation can be produced only once. This effect is highly specific. Structural analogues of 5-HT, such as tryptolines (TLNs) or 5-methoxy-N,N dimethyltryptamine (5-MNDTA), with agonist properties in animal behavioral models, cannot induce a 5-HT-like response in the platelet system; however, they competitively inhibit the platelet aggregation response (PAR) to 5-HT if preincubated with the platelets. Inhibition constants of TLNs for 5-HT and E are of an order of magnitude higher than the K_M values for aggregation.

A transient '5-HT-like' PAR can be produced by compounds structurally unrelated to 5-HT, such as the following: a preparation of adenylylimidodiphosphate (AIP) can cause a transient PAR similar to the 5-HT response with a K_M of 0.2–0.5 mM. Experimental evidence suggests that this ATP analogue is also an inhibitor of the second phase of aggregation (17) and release of dense vesicle content, thereby exhibiting only a monophasic and transient PAR. A '5-HT-like' PAR can also be produced by a combination of E and ATP at concentrations at which neither by itself causes aggregation. These transient PARs may serve as tools to investigate the underlying mechanism of the self-restricted yet exhaustible ability of platelets to respond to the aggregation inducer.

Drug-free schizophrenic patients having their first psychotic episode have a transient 5-HT PAR indistinguishable from that of control subjects. However, after 2–3 weeks on chlorpromazine (CPZ) as the sole neuroleptic, the patients defined as schizophrenic could be divided into 2 sub-groups: a) those who developed an enhanced and irreversible PAR to 5-HT and were reported as good clinical responders; and b) those whose transient 5-HT PAR remained unchanged for 10–12 weeks and were reported as slow, poor or non-responders.

The study of the mechanism of the enhanced 5-HT PAR in CPZ-treated patients may contribute to the understanding of certain aspects of the mode of action of CPZ.

Acknowledgments

The technical help of Miss *Bilha Glantz* is gratefully acknowledged. We also wish to thank Dr. *R. Coleman* for the electron micrographs of platelets and Drs. *E. Klein* and *S. Goldstein* for the blood samples. This work was supported by grants from the Israel Center for Psychobiology, Jerusalem; Research and Development Foundation, Technion, Haifa; and the Wellcome Trust, London.

References

1 Baumgartner, J.R. and Born, G.V.R.: Effects of 5-hydroxytryptamine on platelet aggregation. Nature, Lond. *218:* 137 (1968).
2 Bleuler, E.: Dementia praecox or the group of schizophrenias; Zinkin (trans.), pp. 14–226 (International University Press, New York 1950).
3 Born, G.V.R.; Juengjaroen, K., and Michal, F.: Relative activities on aggregation and uptake by human blood platelets of 5-hydroxytryptamine and several analogues. Br. J. Pharmacol. *44:* 117–139 (1972).
4 Boullin, D.J.; Grahame-Smith, D.G.; Grimes, R.P.J., and Woods, H.F.: Inhibition of 5-hydroxytryptamine-induced human blood platelet aggregation by chlorpromazine and its metabolites. Br. J. Pharmacol. *53:* 121–125 (1975).
5 Boullin, D.J. and Grimes, R.P.J.: Increased platelet aggregation in patients receiving chlorpromazines: responses to 5-hydroxytryptamine, dopamine and N-dimethyl dopamine. Br. J. clin. Pharmacol. *3:* 649–653 (1976).
6 Clement-Cormier, Y.C.; Kebabian, J.W.; Petzold, G.L., and Greengard, P.: Dopamine sensitive adenylate cyclase in mammalian brain: a possible site of action of antipsychotic drugs. Proc. natn. Acad. Sci. USA *71:* 1113–1117 (1974).
7 Feighner, J.P.; Robins, E.; Guze, S.B.; Woodruff, R.A.; Winokur, G., and Munoz, R.: Diagnostic criteria for use in psychiatric research. Archs gen. Psychiat. *26:* 57–63 (1972).
8 Glover, V.; Sandler, M.; Owen, F., and Riley, G.J.: Dopamine is a monoamine oxidase-B substrate in man. Nature, Lond. *265:* 80–81 (1977).

9a Grahame-Smith, D.G.: Inhibitory effect of chlorpromazine on the syndrome of hyperactivity produced by L-tryptophan or 5-methyoxy-N-dimethyltryptamine in rats treated with a monoamine oxidase inhibitor. Br. J. Pharmacol. *43:* 856–864 (1971).

9b Green, A.R.; Heal, D.; Boullin, D.J., and Grahame-Smith, D.G.: Potentiation of hyperactivity produced by monoamine oxidase inhibitor and L-Dopa in rats after repeated treatment with chlorpromazine. Psychopharmacologia *49:* 287–300 (1976).

10 Hefez, A.; Oppenheim, B.; Glantz, B., and Youdim, M.B.H.: Chlorpromazine therapy and platelet aggregation response to serotonin. Israel J. med. Sci. (in press).

11 Hoffman, P.G.; Zinder, O.; Bonner, W.M., and Pollard, H.B.: Role of ATP and β, γ iminoadenosine triphosphate in the stimulation of epinephrine and protein release from isolated adrenal secretory vesicles. Archs Biochem. Biophys. *176:* 375–388 (1976).

12 Holman, R.B.; Seagraves, E.; Elliott, G.R., and Barchas, J.D.: Stereotyped hyperactivity in rats treated with tranylcypromine and specific inhibitors of 5-HT reuptake. Behav. Biol. *16:* 507–514 (1976).

13 Holmsen, H.: Biochemistry of the platelet release reaction; in Elliot and Knight, Biochemistry and pharmacology of platelets; pp. 175–206 (Elsevier, North-Holland 1975).

14 Mills, D.C. and Roberts, G.C.: Membrane active drugs and the aggregation of human blood platelets. Nature, Lond. *213:* 35 (1967).

15 Mitchell, J.R.A. and Sharp, A.A.: Platelet clumping *in vitro*. Br. J. Haemat. *10:* 78–93 (1964).

16 Murphy, D.L. and Donnelly, D.H.: Monoamine oxidase in man: enzyme characteristics in platelets, plasma and other human tissues; in Usdin, Neuropsychopharmacology of monoamines and their regulatory enzymes; pp. 71–86 (Raven Press, New York 1974).

17 Oppenheim, B. and Youdim, M.B.H.: A nucleotide with characteristic platelet aggregation and inhibition properties similar to 5-hydroxytryptamine. Br. J. Pharmacol. *66:* 948 (1979).

18 Oppenheim, B. and Youdim, M.B.H.: Synergism between adrenaline and adenine nucleotides in producing platelet aggregation responses; in 7th Int. Congr. on Thrombosis and Haemostasis, 1979 (in press).

19 Oppenheim, B.; Youdim, M.B.H.; Goldstein, S., and Hefez, A.: Human platelets as a neuronal model for the study of the pharmacological activity of tryptolines and neuroleptics. Israel J. med. Sci. *104:* 1096 (1979).

20 Pletcher, A.: Metabolism, transfer and storage of 5-hydroxytryptamine in blood platelets. Br. J. Pharmac. Chemother. *32:* 1–16 (1968).

21 Pletcher, A.: Platelets as models for monaminergic neurons; in Youdim, Lovenberg, Sharman and Lagnado, Essays in neurochemistry and neuropharmacology; vol. 3, pp. 49–102 (John Wiley, London 1977).

22 Pletcher, A. and Da Prada, M.: The organelles storing 5-hydroxytryptamine in blood platelets; in Elliot and Knight, Biochemistry and pharmacology of platelets; pp. 261–286 (Elsevier, North-Holland 1975).

23 Riederer, P.; Youdim, M.B.H.; Rausch, W.D.; Birkmayer, W.; Jellinger, K., and Seeman, D.: On the mode of action of L-deprenyl in the human central nervous system. J. Neural Transm. *43:* 217–226 (1978).

24 Schneider, K.: Clinical psychopathology; Hamilton (trans.), pp. 95–144 (Grune and Stratton, New York 1959).

25 Seeman, P.: Antischizophrenic drugs – membrane receptor sites of action. Biochem. Pharmacol. *26:* 1741–1748 (1977).

26 Seeman, P. and Lee, T.: Antipsychotic drugs: direct correlation between clinical potency and presynaptic action of dopamine neurons. Science, N.Y. *188:* 1217–1219 (1975).
27 Shoemaker, D.W.; Cummins, J.T., and Bidder, T.G.: Carbolines in rat arcuate nucleus. Neuroscience *3:* 233 (1978).
28 Skaer, R.J.; Peters, P.S., and Emmines, J.P.: The localization of calcium and phosphorus in human platelets. J. Cell Sci. *15:* 679–692 (1974).
29 Youdim, M.B.H.; Grahame-Smith, D.G., and Woods, H.F.: Some properties of human platelet monoamine oxidase in iron deficiency anaemia. Clin. Sci. mol. Med. *50:* 479–485 (1976).
30 Youdim, M.B.H.; Oppenheim, B., and Goldstein, S.: The effect of tryptolines on platelet aggregation and monoamine metabolism; in 7th Int. Congr. on Pharmacology, Paris; p. 72 (1978).

Dr. M.B.H. Youdim, Department of Pharmacology, Faculty of Medicine,
Technion–Israel Institute of Technology, Haifa (Israel)

Prog. biochem. Pharmacol., vol. 16, pp. 133–140 (Karger, Basel 1980)

Concentrations of Monoamine Metabolites and Chlorpromazine in Cerebrospinal Fluid for Prediction of Therapeutic Response in Psychotic Patients Treated with Neuroleptic Drugs

Göran Sedvall

Laboratory of Experimental Psychiatry, Department of Psychiatry (Karolinska Hospital), Karolinska Institute, Stockholm

Introduction

The schizophrenias represent a major human and social health problem. The incidence of these disorders in the population of western societies is of the order of 1%. The nature and causes of the schizophrenias have been the subject of psychiatric research for more than a century. It is now generally accepted that at least some forms of schizophrenia have a marked familial disposition (16) which has been shown to be related to hereditary factors (8). It is also widely acknowledged that the symptoms of schizophrenia can be alleviated by treatment with drugs that block catecholamine receptors in the central nervous system (7).

The demonstration that antipsychotic and psychotomimetic drugs interfere with monoamine receptors in the brain has focussed interest on monoaminergic transmitter mechanisms in the pathophysiology of schizophrenia (9). Measurement of the concentrations of the major monoamine metabolites in lumbar cerebrospinal fluid (CSF) of patients has been used extensively during recent years as a tool for examining transmitter turnover in central monoamine neurons. The steady-state concentrations of the major monoamine metabolites, 5-hydroxyindoleacetic acid (5-HIAA), homovanillic acid (HVA) and 3-methoxy-4-hydroxyphenylethylene glycol (MOPEG) as measured by highly specific mass fragmentographic methods (2, 4, 15) have been shown to reliably reflect changes in central monoamine turnover during treatment with psychoactive drugs (3, 10, 12).

The introduction of neuroleptic drugs for the treatment of schizophrenia in 1952 resulted in a marked change in the approach to the treatment of acute psychotic reactions. However, during recent years it has become increasingly evident that neuroleptic treatment results in a number of side effects including sedation, extrapyramidal manifestations and tardive dyskinesia. This calls for an increased effort to find new forms of treatment in schizophrenia. Another major problem in the treatment of psychosis is the large variation in drug response to conventional antipsychotic drugs. This variation is due to three major factors: 1) biochemical heterogeneity within the group of the schizophrenias; 2) individual variation in the rate of drug metabolism; 3) individual variation in receptor sensitivity to the neuroleptic drugs.

In the present communication we give examples of these causes of variance and indicate how this variation can be minimized by using chemical methods to control variation in drug response.

Results and Discussion

Evidence for a Biochemical Heterogeneity of Schizophrenic Syndromes

Previous studies indicate that schizophrenic patients show great variation in the concentration of 5-HIAA in the CSF (11). In recent studies on healthy volunteers and schizophrenic patients we were interested in analysing some of the factors responsible for this variation. In groups of schizophrenic patients we found highly significant relationships ($p < 0.001$) between deviant 5-HIAA concentration in the CSF and family history of the disorder (14). Most patients with schizophrenia in the family had a high 5-HIAA concentration in the CSF (fig. 1). Patients who did not have schizophrenia in the family had a fairly normal 5-HIAA distribution in the CSF. These results indicate the existence of two forms of schizophrenia. One form seems to have a familial disposition for the disorder and a disturbed central serotonin metabolism. The other has a normal metabolism of serotonin in the brain and does not seem to be related to hereditary mechanisms. We believe that these results are important for future studies aimed at analysing the mechanism of action of antipsychotic drugs in patients. It seems likely that patients with familial disposition, who have specific alterations of central monoamine metabolism, should show a specific response to certain types of antipsychotic drugs. Few systematic studies have probed the question of whether family forms of schizophrenia have a specific response to antipsychotic drug treatment. The present results should be a further stimulus for this type of study.

Relationships between Clinical and Biochemical Effects and Drug Concentrations in Body Fluids

In chlorpromazine-treated patients the concentrations of the unaltered drug and two of its active metabolites, 1-desmethylchlorpromazine and 7-hydroxychlorpromazine, were determined in the plasma. Mass fragmentographic methods were used for the analysis (1). After drug treatment for 2 weeks a significant correlation was found between the chlorpromazine concentrations in plasma and CSF and the therapeutic effect (fig. 2). Although this correlation between drug

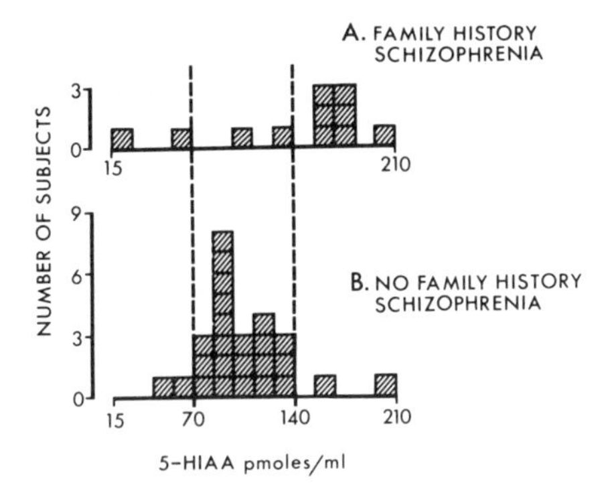

Fig. 1. Distribution of 5-HIAA in CSF of untreated schizophrenic patients in relation to family history of the disorder (14).

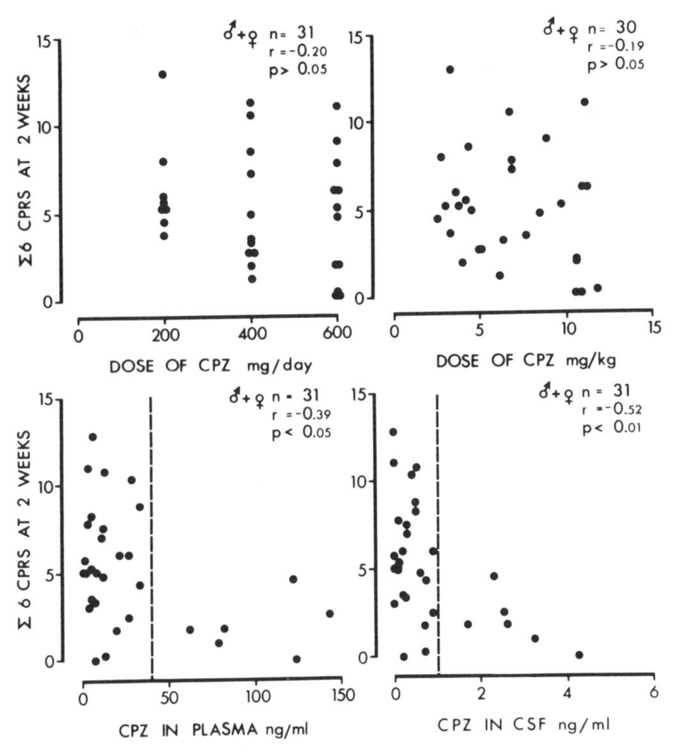

Fig. 2. Relationships between pharmacokinetic data and final morbidity score in schizophrenic patients treated with chlorpromazine for 2 weeks. Patients were treated with fixed doses of chlorpromazine (200–600 mg/day). The doses were given on a blind basis (18).

concentration and therapeutic response was not maintained after 4 weeks' treatment, on this occasion a significant correlation was still found between the drug concentration and side effects like extrapyramidal symptoms and sedation. The disappearance of the correlation between antipsychotic effect and drug concentration between 2 and 4 weeks may be related to the induction of secondary effects unrelated to drug concentration during long-term treatment. A similar change in the relationship to drug concentration was present for HVA. Thus after 2 weeks' treatment there was a significant correlation between the chlorpromazine concentration in the plasma and the HVA concentration in the CSF. After 4 weeks' treatment this correlation too had disappeared (13).

These results support the view that inductive biochemical changes take place during long-term treatment with chlorpromazine. Such an induction seems to occur for both the antipsychotic action and the central monoamine metabolism. The results further emphasise the need to perform long-term studies in experimental animals and patients in order to elucidate the mechanism of action of antipsychotic drugs. The results also indicate that drug concentration-effect relationships may be different at various time intervals after the beginning of pharmacotherapy.

Long-Term Effects of Neuroleptic Drug Treatment on Concentrations of Monoamine Metabolites in the Cerebrospinal Fluid

In order to examine how different types of antipsychotic drugs affect central monoamine metabolism in patients the following type of study was designed (5, 19): Schizophrenic patients who had been drug free for at least 2 weeks before the study were given a standardised drug treatment that was maintained at a constant level for 4 weeks. Before 2 weeks and after 4 weeks of treatment lumbar CSF and blood plasma were sampled. Psychotic signs and symptoms and side effects were clinically rated on the same day as CSF samples were withdrawn. Concentrations of the major monoamine metabolites HVA, MOPEG and 5-HIAA were determined by mass fragmentographic techniques (17).

Different types of phenothiazine, thioxanthene, butyrophenone and benzamide neuroleptics were examined. All the drugs were administered in constant doses considered to induce an optimal antipsychotic action. Table I demonstrates that all the drugs with the exception of clozapine caused a significant elevation of the HVA concentration in CSF. All the drugs elevating HVA caused about the same elevation after 2 and 4 weeks' treatment, indicating that there was no significant tolerance to the HVA elevating effect for any of the drugs. There were marked quantitative differences between the effects of the different neuroleptics. Thus thiothixene was significantly more potent in elevating HVA than melperone, and clozapine caused no elevation at all.

The quantitative relationship between elevation of HVA concentration and antipsychotic effect was studied in patients treated with chlorpromazine, thio-

Table I. Effect of drug treatment on monoamine metabolite levels in cerebrospinal fluid from psychotic patients

Drug	n	Dose mg/day	HVA (pmoles/ml)		MOPEG (pmoles/ml)	
			before	during	before	during
Chlorpromazine	8	200–600	221 ± 23	327 ± 44[1]	41 ± 2	25 ± 1[2]
Thiothixene	15	30	235 ± 22	381 ± 35[2]	41 ± 4	41 ± 3
Melperone	13	300	218 ± 28	254 ± 31[2]	42 ± 4	35 ± 4[2]
Sulpiride	6	800	219 ± 54	320 ± 51[1]	41 ± 3	39 ± 2
Clozapine	7	600	251 ± 51	228 ± 49	42 ± 3	43 ± 2

Patients were treated for 2 to 4 weeks.
[1] $p < 0.01$.
[2] $p < 0.001$.

thixene and melperone. Chlorpromazine-treated patients showed a positive relationship between the elevation of the HVA concentration and the antipsychotic effect after 2 but not after 4 weeks of treatment (13). For neither thiothixene nor melperone were such relationships obtained (18). In all the drugs examined the antipsychotic effect developed slowly and was more marked after 4 than 2 weeks' treatment, but about the same HVA elevation was obtained at 2 weeks as at 4 weeks. These results contradict the view that a close quantitative relationship exists between interaction with central dopaminergic mechanisms and antipsychotic effect in the series of antipsychotic drugs. The lack of such a relationship could be explained by several factors. It seems possible that the neuroleptic drugs can induce the antipsychotic effect by a multitude of biochemical effects which act synergistically. It is also possible that interaction with dopaminergic mechanisms in specific brain regions explains the lack of correlation found between HVA elevation in CSF and clinical response in the present study. However, the latter explanation does not seem likely since all antipsychotic drugs examined so far seem to induce the greatest HVA elevation in striatum which supplies the major amount of HVA to CSF (7).

Some of the antipsychotic drugs also affected the concentrations of noradrenaline and serotonin metabolites in the CSF. Thus chlorpromazine treatment resulted in a marked reduction of the MOPEG concentration in CSF, and the 5-HIAA concentration was also reduced (19). The mechanisms for the latter effects have not been elucidated, but they seem to be related to specific effects of chlorpromazine since thiothixene, haloperidol and sulpiride did not induce such effects. Melperone had an interesting pattern of action. There was no alteration in the MOPEG concentration after 2 weeks of treatment. However, a significant reduction was obtained after 4 weeks when the maximal antipsycho-

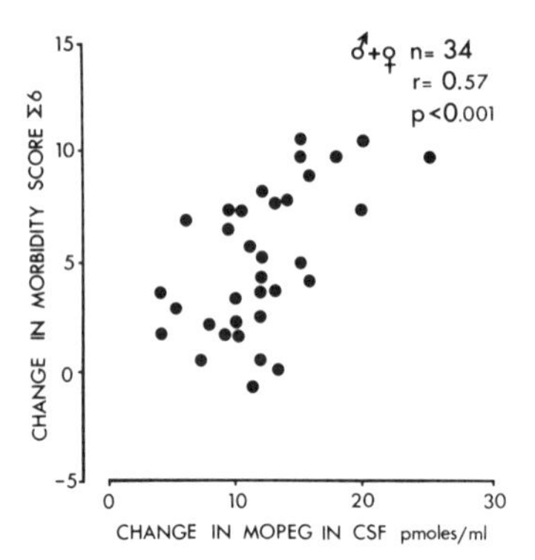

Fig. 3. Relationship between change in MOPEG level in CSF and morbidity score using the CPRS scale after 4 weeks of chlorpromazine treatment. Patients were treated with a fixed dose of chlorpromazine (200–600 mg/day).

tic effect was also present. These results demonstrate that different types of neuroleptic drugs induce specific spectra of effects on the monoamine metabolite concentrations in CSF. The time course for the different effects seems to vary from drug to drug, and it is of special interest to note that after treatment with both chlorpromazine and thiothixene there was a significant correlation between the reduction in MOPEG concentration and the therapeutic response after 4 weeks of treatment (fig. 3) (6). None of these drugs exhibited a significant correlation between the HVA elevation and the therapeutic response at this late time point.

The results described indicate that as in animal experiments most neuroleptic drugs induce a marked elevation of the major dopamine metabolite HVA in the brain of drug-treated patients. However, there appears to be no direct relationship between this effect and antipsychotic potency. The group of antipsychotic drugs induce a spectrum of other biochemical effects related to changes of central monoaminergic mechanisms. Some of these changes are quantitatively related to the antipsychotic effects in the later phases of treatment as well. The results indicate that inductive changes related to alteration of central noradrenergic and serotonergic mechanisms may also be involved in the mechanism of action of some antipsychotic drugs.

Summary

Concentrations of the major monoaminergic transmitter metabolites HVA, MOPEG and 5-HIAA were determined in the cerebrospinal fluid (CSF) of untreated schizophrenic patients. Patients with aberrated concentrations of 5-HIAA and HVA in CSF had schizophrenia in their families in a frequency significantly higher than that of patients with normal concentrations. These results may indicate the existence of a subgroup of schizophrenic patients having a family disposition for the disorder and an aberrated transmission from central serotonin and dopamine neurons.

In young healthy volunteers, aberrated monoamine metabolite concentrations in CSF were also related significantly to a history of psychiatric morbidity in the family.

In psychotic patients treated with chlorpromazine significant correlations were obtained between therapeutic outcome on the one hand, and both biochemical effects related to central monoamine metabolism and chlorpromazine concentrations in the CSF on the other hand. Patients with chlorpromazine concentrations above 1 ng/ml in CSF or 40 ng/ml in plasma responded more favourably than patients with concentrations below those levels.

The results indicate that biochemical and pharmacokinetic data may be of value for diagnostic classification and prediction of therapeutic outcome in drug-treated schizophrenic patients.

Acknowledgements

The present studies were supported by grants from the Swedish Medical Research Council (No. 21X-03560), the National Institute of Mental Health, Bethesda, Maryland, USA (MH 27254), F. Hoffmann-La Roche & Co., Basel, Switzerland, Magnus Bergvalls Stiftelse and Karolinska Institutet.

References

1 Alfredsson, G.; Wode-Helgodt, B., and Sedvall, G.: A mass fragmentographic method for the determination of chlorpromazine and two of its active metabolites in human plasma and CSF. Psychopharmacology *48:* 123–131 (1976).
2 Bertilsson, L.: Quantitative determination of 4-hydroxy-3-methoxyphenyl glycol and its conjugates in cerebrospinal fluid by mass fragmentography. J. Chromatogr. *87:* 147–153 (1973).
3 Bertilsson, L.; Åsberg, M., and Thoren, P.: Differential effect of chlorimipramine and nortriptyline on cerebrospinal fluid metabolites of serotonin and noradrenaline in depression. Eur. J. clin. Pharmacol. *7:* 365–368 (1974).
4 Bertilsson, L.; Atkinson, A.J.; Althaus, J.R.; Härfast, Å.; Lindgren, J.-E., and Holmstedt, B.: Quantitative determination of 5-hydroxyindole-3-acetic acid in cerebrospinal fluid by gas chromatography-mass spectrometry. Analyt. Chem. *44:* 1434–1438 (1972).
5 Bjerkenstedt, L.; Gullberg, B.; Härnryd, C., and Sedvall, G.: Monoamine metabolite levels in cerebrospinal fluid of psychotic women treated with melperone or thiothixene. Arch. Psychiat. NervKrankh. *224:* 107–118 (1977).
6 Bjerkenstedt, L.; Gullberg, B.; Härnryd, C., and Sedvall, G.: Relationships between clinical and biochemical effects of melperone and thiothixene in psychotic women. Arch. Psychiat. NervKrankh. *227:* 181–192 (1979).
7 Carlsson, A.: Antipsychotic drugs, neurotransmitters and schizophrenia. Am. J. Psychiat. *135:* 164–173 (1978).
8 Kety, S.S.; Rosenthal, D.; Wender, P.H.; Schulsinger, F., and Jacobsen, B.: Mental illness in the biological and adoptive families of adopted individuals who have become

schizophrenic: A preliminary report based upon psychiatric interviews; in Fieve, Rosenthal and Brill, Genetic research in psychiatry; pp. 147–165 (John Hopkins University Press, Baltimore and London 1975).
9 Matthysse, S. and Lipinski, J.: Biochemical aspects of schizophrenia. Ann. Rev. Med. *26:* 551–565 (1975).
10 Sedvall, G.; Alfredsson, G.; Bjerkenstedt, L.; Eneroth, P.; Fyrö, B.; Härnryd, C.; Swahn, C.-G.; Wiesel, F.-A., and Wode-Helgodt, B.: Selective effects of psychoactive drugs on levels of monoamine metabolites and prolactin in cerebrospinal fluid of psychiatric patients; in Proc. 6th Congr. of Pharmacology, vol. 3, pp. 255–267 (Helsinki 1975).
11 Sedvall, G.; Bjerkenstedt, L.; Swahn, C.-G.; Wiesel, F.-A., and Wode-Helgodt, B.: Mass fragmentography to study dopamine metabolism; in Costa and Gessa, Advances in biochemical psychopharmacology; vol. 16, pp. 343–348 (Raven Press, New York 1977).
12 Sedvall, G.; Fyrö, B.; Nybäck, H.; Wiesel, F.-A., and Wode-Helgodt, B.: Mass fragmentometric determination of homovanillic acid in lumbar cerebrospinal fluid of schizophrenic patients during treatment with antipsychotic drugs. J. psychiat. Res. *11:* 75–80 (1974).
13 Sedvall, G. and Grimm, V.E.: The cerebrospinal fluid and plasma as tools for obtaining biochemical and pharmacokinetic data in neuroleptic therapy; in Burrows and Norman, Plasma level measurements of psychotropic drugs and clinical response (Marcel Dekker, New York) (in press).
14 Sedvall, G. and Wode-Helgodt, B.: Relationships in schizophrenic patients between aberrant monoamine metabolite concentrations in cerebrospinal fluid and family history of the disorder. Nature, Lond. (in press).
15 Sjöquist, B. and Änggård, E.: Gas chromatographic determination of homovanillic acid in human cerebrospinal fluid by electron capture detection and by mass fragmentography with a deuterated internal standard. Analyt. Chem. *44:* 2297–2301 (1972).
16 Slater, E. and Cowie, V.A.: The genetics of mental disorders (Oxford University Press, London 1971).
17 Swahn, C.-G.; Sandgärde, B.; Wiesel, F.-A., and Sedvall, G.: Simultaneous determination of the three major monoamine metabolites in brain tissue and body fluids by a mass fragmentographic method. Psychopharmacology *48:* 147–152 (1976).
18 Wode-Helgodt, B.; Borg, S.; Fyrö, B., and Sedvall, G.: Clinical effects and drug concentrations in plasma and cerebrospinal fluid in psychotic patients treated with fixed doses of chlorpromazine. Acta psychiat. scand. *58:* 149–173 (1978).
19 Wode-Helgodt, B.; Fyrö, B.; Gullberg, B., and Sedvall, G.: Effect of chlorpromazine treatment on monoamine metabolite levels in cerebrospinal fluid of psychotic patients. Acta psychiat. scand. *56:* 129–142 (1977).

Dr. Göran Sedvall, Laboratory of Experimental Psychiatry, Department of Psychiatry (Karolinska Hospital), Karolinska Institute, S–104 01 Stockholm (Sweden)

Prog. biochem. Pharmacol., vol. 16, pp. 141–154 (Karger, Basel 1980)

Purinergic Nerves and Receptors

G. Burnstock

Department of Anatomy and Embryology, University College London, London

Introduction

There is a growing recognition of the potent extracellular actions of ATP and related nucleotides and nucleosides on excitable membranes and that these may be involved in physiological regulatory mechanisms (see 4, 5, 6, 9, 15, 26). The high sensitivity of smooth and cardiac muscle cell membranes to purine nucleotides and nucleosides was first reported in 1929 in a study of the pharmacology of nucleic acid derivatives (17). *Holton* (19) suggested that ATP was released during antidromic stimulation of sensory nerves, and it has been proposed that adenosine (or ATP) is a physiological regulator of blood flow in coronary, renal, skeletal muscle and cerebral vascular beds (see 2, 10).

The 'purinergic nerve hypothesis' was first put forward in 1972, when evidence was presented that a purine nucleotide, probably ATP, was the principal transmitter released from the non-adrenergic, non-cholinergic nerves supplying the gastrointestinal tract, urinary bladder and lung (4). A tentative model of the mechanisms of storage, release and inactivation of ATP from purinergic nerves was proposed (fig. 1). A considerable body of evidence has now accumulated in support of the hypothesis and the presence of purinergic nerves has been proposed in a variety of organs, including the gut, lung, trachea, seminal vesicles, urinary bladder, oesophagus, eye and probably parts of the cardiovascular and central nervous systems (5, 6, 9). Knowledge of purinergic neurons is in its infancy when compared to the large body of literature on adrenergic and cholinergic nerves but certain characteristics are established (see 4, 5, 9 for details).

Inhibitory junction potentials characterised by long latencies (40–80 msec) and duration (800–1,200 msec) have been recorded in single smooth muscle cells in response to stimulation of enteric nerves in the circular and longitudinal muscle coats of some regions of the gut. These persist in the presence of

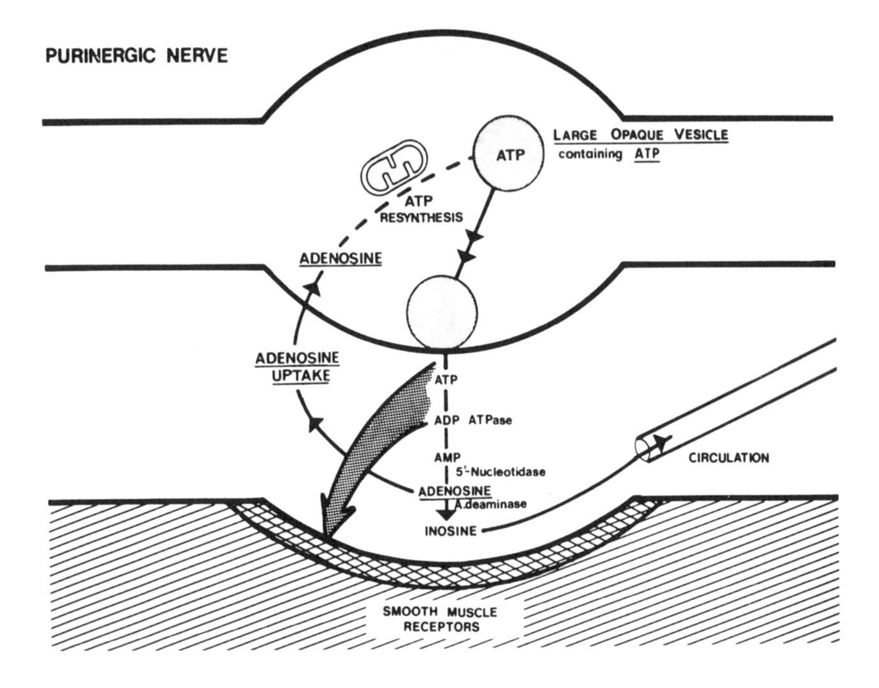

Fig. 1. Schematic representation of synthesis, storage, release and inactivation of autonomic transmitters at purinergic junctions (from Ref. 4).

adrenergic and cholinergic blocking agents or after degeneration of sympathetic adrenergic nerves but are abolished when nerve conduction is blocked by tetrodotoxin. Repetitive stimulation of purinergic nerves results in summation of individual inhibitory junction potentials and hyperpolarisations of up to 50 mV. The response to repetitive nerve stimulation shows a rapid decay and on cessation of stimulation there is a 'rebound contraction.'

Tissues in which purinergic nerves have been demonstrated experimentally contain a predominance of 'large opaque vesicles,' which can be distinguished from the 'large granular vesicles' found in small numbers in both adrenergic and cholinergic nerves (4). Large opaque vesicles are larger (80–200 nm) than the large granular vesicles (60–120 nm), have a less prominent halo between the granular core and the vesicle membrane and have a less granular matrix. They are unaffected by 6-hydroxydopamine, a drug which produces short-term 'loading' or long-term 'destruction' of adrenergic nerves, or by surgical removal of adrenergic nerves.

The physiological roles of purinergic nerves are beginning to be understood (see 5, 9), especially in the gastrointestinal tract where they are involved in inhibitory reflexes which facilitate passage of material through the alimentary canal by opening sphincters, increasing stomach size, and expanding the intestine in front of an advancing bolus (13). Non-cholinergic, non-adrenergic inhibitory nerves comparable to those found in the gastrointestinal tract have been identified in the trachea (16); these may be involved in the cough mechanism. Purinergic nerves have been implicated in bronchodilation in the vertebrate lung and in the vasodilator control of some blood vessels (see 5). Adenyl compounds are also thought to play a role in the central nervous system (6), since electrical stimulation of different areas of brain leads to release of nucleotides, and either electrophysiological or biochemical changes have been detected when adenosine or ATP is applied to various regions of the brain. Certain clinical observations implicate adenine nucleotides in schizophrenia and psychotic and neurotic depression.

Evidence that ATP is a Transmitter

In order for a substance to be regarded as a neurotransmitter there are five criteria that should be satisfied, namely: (a) synthesis and storage of transmitter in nerve terminals; (b) release of transmitter during nerve stimulation; (c) postjunctional responses to exogenous transmitter that mimic responses to nerve stimulation; (d) enzymes that inactivate the transmitter and/or an uptake system for the transmitter or its breakdown products; (e) drugs that produce parallel blocking or potentiating effects on the responses to both exogenous transmitter and nerve stimulation.

In experiments aimed at identifying the transmitter in non-adrenergic, non-cholinergic inhibitory nerves supplying gastrointestinal smooth muscle, ATP was found to fulfil these criteria (see 4, 5, 9). Other substances, including 3′, 5′-catecholamines, 5-hydroxytryptamine, cyclic adenosine monophosphate (cAMP), histamine, prostaglandins, various amino acids such as alanine, arginine, histidine, glycine, glutamic acid and γ-aminobutyric acid (GABA) and the polypeptides, enkephalin, neurotensin, vasoactive intestinal peptide (VIP), somatostatin, bradykinin and substance P, have been rejected as contenders for this role by most workers, on the grounds that they were either inactive or did not mimic the nerve-mediated responses, that specific blocking drugs for these substances did not affect the nerve-mediated response or that their action was by stimulation of nerves and not by direct action on smooth muscle.

Storage

Both ATP and the enzyme systems that synthesize ATP occur ubiquitously in cells, so that it is not contentious that non-adrenergic, non-cholinergic inhibitory nerves are also able to produce and store ATP. Tritiated-adenosine is taken up by preparations of stomach and intestine and most of it is rapidly converted into (^{3}H)ATP. Analysis of the radioactivity in serial frozen sections has shown that when the tissue is incubated in low concentrations of ^{3}H-adenosine, most of the label is stored as ^{3}H-ATP in nerves.

It is likely that ATP is stored in the large opaque vesicles (LOV) that are predominant in nerves since studies in our laboratory suggest that after exposure of intestine to low concentrations of ^{3}H-adenosine for short periods (45 sec) there is selective accumulation of silver grains over nerve profiles containing LOV. Furthermore, patients infected with *Trypanosoma cruzi,* a unicellular parasite responsible for Chagas' disease, have damaged LOV in nerve profiles of their intestines. This is of particular interest in view of the fact that this parasite is unable to synthesise its own adenine.

Quinacrine is a useful tool for detecting purinergic nerves since it is known to bind to ATP and can be localised by a fluorescence histochemical method (23). It gives positive staining of adrenal medullary cells, megakaryocytes and blood platelets, all known to contain high levels of ATP. Microsomal fractions obtained by differential and sucrose-density gradient centrifugation of homogenates of purinergically innervated preparations (taenia coli and urinary bladder) preloaded with ^{3}H-adenosine and ^{14}C-quinacrine showed peaks of ^{14}C which corresponded to peaks of ^{3}H-ATP. Quinacrine-positive nerve cell bodies and varicose fibres have been demonstrated in gut and also in gall bladder, urinary bladder and portal vein, where pharmacological evidence for purinergic transmission has been presented, but are absent from the iris, which contains abundant adrenergic and cholinergic but not purinergic nerves (9).

Release

There was early evidence that ATP was released during stimulation of non-adrenergic, non-cholinergic nerves. For example, adenosine and inosine were found in the venous efflux from the stomachs of guinea-pigs and toads on stimulation of the vagus nerves. These compounds are likely to be the break-down products of adenine nucleotides, because ATP introduced into the same perfusing system was quickly broken down into comparable proportions of adenosine and inosine: furthermore, stimulation of the intramural nerves resulted in the release of radioactive compounds from guinea-pig taenia coli which had previously been incubated in ^{3}H-adenosine. This release was blocked by tetrodotoxin. It is unlikely that the release of nucleosides was due to stimulation of cholinergic fibres rather than of non-adrenergic inhibitory fibres in the vagus nerves: stimulation of the vagal roots (which contain preganglionic parasympathetic fibres in the vagus nerves making synaptic connections with non-adrenergic inhibitory postganglionic neurons in the wall of the stomach) resulted in increased nucleoside efflux, while stimulation of the cervical sympathetic branch of the vagus nerves (which contain cholinergic fibres) did not.

The possibility that the purine nucleotides or nucleosides released from nerves are not neurotransmitter substances but are released from nerve membranes during propagation of an action potential has been considered. However, the amount of nucleosides collected during stimulation of non-adrenergic, non-cholinergic inhibitory nerves was calculated to be at least 1,000-fold greater than that released as a direct result of the process of axon membrane activation during impulse propagation.

The possibility that ATP is released as a result of antidromic stimulation of enteric sensory nerves was also ruled out when it was shown that the non-adrenergic inhibitory response to stimulation of the vagal nerves supplying the rabbit stomach was abolished after degeneration of the efferent but not the afferent component achieved by vagal section above the nodose ganglion eight days previously.

The problem of whether ATP released during stimulation of purinergic nerves comes from nerve or secondarily from muscle has been resolved recently in our laboratory (11). It was shown that while there was a two- to six-fold increase in ATP release from the guinea-pig taenia coli or urinary bladder during isometric responses to purinergic nerve stimulation, there was no significant release of ATP during comparable responses elicited by direct muscle stimulation. Further, the nerve-mediated release of ATP was Ca^{++} dependent; a 10-fold reduction in Ca^{++} concentration resulted in an 80–90% reduction in both the mechanical response and the ATP release (11). This finding is consistent with that of *Holman and Weinrich* (18), who showed that the amplitude of the inhibitory junction potentials in the taenia coli was reduced by 80% when the external Ca^{++} was reduced 10-fold. Release from muscle was also considered unlikely, since stimulation of portions of Auerbach's plexus from turkey gizzard (heavily innervated by non-adrenergic inhibitory nerves), dissected free of the underlying muscle, still resulted in efflux of purine nucleotides, in this case mostly AMP.

Receptor Activation

The form and time course of the response to exogenously applied ATP closely mimics that of the response to non-adrenergic, non-cholinergic inhibitory nerve stimulation. Typically, the relaxations of the gut produced by ATP and nerve stimulation rapidly reach a maximum that declines quickly; this is in contrast to the relaxations produced by noradrenaline and sympathetic nerve stimulation, which reach a maximum more slowly and are maintained for a longer time. It is unlikely that ATP causes relaxation by initiating action potentials in non-adrenergic inhibitory nerves since tetrodotoxin abolished the inhibitory responses of atropinised intestinal muscle to stimulation of both perivascular and intramural nerves but did not affect the relaxation produced by ATP.

In the intestine the most potent inhibitory purine compounds are ATP and ADP, which are about equipotent, with a threshold concentration for relaxation of the taenia coli of about 10^{-7} M. AMP and adenosine have about 1/100 the potency of ATP. The following related compounds produce no effects, even with concentrations as high as 10^{-3} M: the purine base, adenine; the deaminated nucleoside, inosine, and its mononucleotide, IMP; the pyrimidine nucleotide, uridine, and its mononucleotide UMP. The sensitivity of the gut to cAMP is also low. ATP is more likely to be the 'purinergic' neurotransmitter since ^{3}H-adenosine taken up into the taenia coli is rapidly converted and stored mostly as ^{3}H-ATP, with only traces of ^{3}H-ADP detectable.

ATP has been found to be a potent inducer of prostaglandin synthesis (22). This finding, linked with the discovery that the 'rebound contractions' following the inhibitory responses of guinea-pig taenia coli to purinergic nerve stimulation and ATP were blocked by the prostaglandin synthesis inhibitor, indomethacin, led to the suggestion that ATP released from purinergic nerves may be linked with prostaglandins in the control of peristalsis (12).

Inactivation

The rapid recovery of smooth muscle after application of ATP or stimulation of non-adrenergic, non-cholinergic inhibitory nerves and the absence of long-lasting action despite continued stimulation indicate an efficient inactivation mechanism. By analogy with other neuroeffector systems, it has been argued that if the non-adrenergic, non-cholinergic inhibitory nerves act on the gut by releasing ATP, the action of ATP would be terminated by uptake into nerves or smooth mucles and/or breakdown of ATP by enzymes into compounds with greatly reduced potency.

When ATP is added to a perfusion fluid recycled through the vasculature of the stomach, very little ATP is recovered, but the perfusate contains substantially increased amounts of adenosine and inosine as well as some ADP and AMP. While there is no direct evidence for the breakdown of ATP released from nerves by enzymic activity, the gut is known to contain high levels of 5′-nucleotidase and adenosine deaminase. Mg^{++}-activated ATPase localisation has been described in micropinocytotic vesicles in smooth muscle membranes closely adjacent (20 nm) to non-adrenergic, non-cholinergic nerve profiles in the intestine.

An uptake mechanism for adenosine, but not ADP and ATP, has been demonstrated which suggests that ATP released from nerves is broken down to adenosine before uptake occurs, in a manner comparable to uptake of choline after breakdown of acetylcholine released from cholinergic nerves. When the adenosine moiety of ATP is labelled with tritium and the phosphate moiety labelled with ^{32}P, the rate of uptake of ^{3}H into taenia coli is considerably greater than that of ^{32}P. Further support comes from studies of transmitter overflow at different stimulation frequencies. Stimulation at low physiological frequencies

(5 Hz) shows little overflow, while stimulation at 30 Hz leads to substantial overflow. In the presence of low concentrations of dipyridamole, which inhibits adenosine uptake, overflow at low stimulation frequencies is comparable to that at high frequencies.

Drugs

Several groups of drugs are known to block the responses to either adenine nucleotides or nucleosides in a wide variety of preparations (see 7–9). These include anti-malarial drugs such as mepacrine, quinine and quinidine, and methylxanthines such as caffeine, theophylline and aminophylline. Quinidine blocks non-adrenergic, non-cholinergic nerve-mediated responses in the gut and bladder, but only in concentrations that give little confidence in its specificity. It has been shown that the methylxanthines are competitive antagonists to adenosine and AMP, while several new agents which block responses of intestine to ATP have been found. These include the 2-substituted imidazoline compounds such as antazoline and phentolamine and 2-2′-pyridylisatogen. Unfortunately, these drugs are not specific for ATP; however, it has been claimed recently that very low concentrations (10^{-8} M) of apamin, a constituent of bee venom, selectively block both the inhibitory junction potentials in response to purinergic nerve stimulation in the intestine and the hyperpolarisations produced by ATP (31).

Dipyridamole potentiates the responses of the intestine to both ATP and stimulation of non-adrenergic inhibitory nerves by blocking uptake of adenosine (see 9).

Purinergic Receptors

Whilst there are many accounts of adrenergic and cholinergic receptors, comparatively little is known about purinergic receptors (see 8, 10). Studies on the relative potencies of nucleotides and nucleosides on different tissues have shown that ATP and ADP are generally more potent than AMP, cAMP and adenosine (see above). Although the predominant response of most smooth muscles to ATP is relaxation, excitatory responses sometimes occur, for example, in some blood vessels and in the urinary bladder and intestine of lower vertebrates (10).

In intestinal smooth muscle the hyperpolarisation produced by exogenously applied ATP or by purinergic nerve stimulation is due to a specific increase in K^+ conductance. There is evidence to suggest that adenosine acts on adenylate cyclase receptors in neuroblastoma cells, cardiac muscle, and neurons in the brain (see 10). This action leads to production of cAMP, which may then, as a second messenger, modulate the availability of intracellular Ca^{++}, thereby leading

Table I. Criteria for distinguishing two types of purinergic receptor

Antagonists	Agonist potencies	Changes in cAMP	Induction of prostaglandin synthesis
P_1 Methylxanthines	AD ⩾ AMP > ADP ⩾ ATP	Yes	No
P_2 Quinidine imidazolines 2,2'-pyridylisatogen Apamin	ATP ⩾ ADP > AMP ⩾ AD	No	Yes

to changes in cellular activity. It has been claimed, however, that cAMP is not involved in the relaxation of visceral smooth muscle or in the vasodilatation caused by adenosine (10). *Turnheim et al.* (30) concluded from their experiments that adenosine produced dilatation of dog coronary arteries by directly decreasing the Ca^{++} permeability of vascular smooth muscle. The vasodilator action of adenosine is probably not brought about by either β-adrenergic stimulation or by α-adrenergic inhibition. It has also been suggested that various vasodilators may act by means of a common reaction involving critical SH groups, either as part of the receptor complex or as part of the contractile apparatus (21).

Evidence has recently been presented to support the view that purinergic receptors can be separated into two types (8) in a manner analogous to the established classifications of adrenergic receptors into α- and β-adrenoceptors, or cholinergic receptors into muscarinic and nicotinic receptors, and of histamine receptors into H_1 and H_2 receptors. Examination of the various actions of adenine nucleotides and nucleosides and the agents that have been shown to block these actions has led to the suggestion that it is possible to separate purinergic receptors into P_1 and P_2 purinoceptors according to four criteria: relative potencies of agonists, competitive antagonists, changes in levels of cAMP and induction of prostaglandin synthesis (table I).

P_1 purinoceptors: (a) are blocked by methylxanthines (e.g. theophylline, caffeine, and aminophylline); (b) are most sensitive to adenosine and progressively less sensitive to AMP, ADP and ATP; (c) their occupation of an adenylate cyclase component of these receptors leads to changes in cAMP accumulation, and (d) their occupation of the receptor does not lead to prostaglandin synthesis.

P_2 purinoceptors: (a) are blocked by quinidine, 2-substituted imidazolines, 2,2'-pyridylisatogen and apamin; (b) are most sensitive to ATP and progressively

Table II. Effects of P_1 and P_2 antagonists on the actions of ATP and adenosine on the guinea-pig taenia coli and bladder

	Purinergic nerve stimulation	ATP	Adenosine
Taenia coli	Relaxation	Relaxation	Relaxation
P_1 Antagonist	–	–	Block
P_2 Antagonist	Block	Block	–
Urinary bladder	Contraction	Contraction	Relaxation
P_1 Antagonist	–	–	Block
P_2 Antagonist	Block	Block	–

less sensitive to ADP, AMP, and adenosine; (c) their occupation does not lead to changes in cAMP accumulation, and (d) their occupation of the receptor leads to prostaglandin synthesis.

Evidence that the blocking agents act competitively is important to this hypothesis, and it has been claimed that the action of theophylline in antagonising adenosine is competitive in the guinea-pig cerebral cortex, in the human and guinea-pig atrium, in the rabbit and guinea-pig ileum, in the cholinergic terminals in guinea-pig ileum, guinea-pig trachea, and guinea-pig kidney vessels, and in human astrocytoma cells (8). The grounds for this claim are that: (a) the antagonism is readily reversible at all concentrations; (b) raising the concentration of adenosine overcomes the antagonistic action of theophylline; (c) the slopes of adenosine concentration-response curves in the presence of theophylline are not significantly different from the slopes of controls, that is, the slopes are shifted in parallel to the right. The pA_2 values for theophylline, when calculated by the *Arunlakshana and Schild* method (1), were lower than those obtained for the established competitive antagonists of acetylcholinic (muscarinic), catecholaminic, and histaminic receptors. The slope of the plots, however, was close to unity, which is the theoretical value taken to indicate competitive antagonism. Theophylline is also known to inhibit phosphodiesterase, but only with concentrations higher than those required to block adenosine. The drugs that have been found to block the actions of ATP on P_2 purinoceptors are not competitive, although they reportedly have some selective action (28).

In addition to the P_1, P_2 classification of purinergic receptors, it is possible that there are two adenosine receptor sites in a variety of preparations (e.g. liver, adrenal tumours, thyroid, platelets): the 'R' site, occupancy of which leads to activation of adenylate cyclase and which requires the integrity of the ribose ring for activation; and the 'P' site, which mediates inhibition of adenylate cyclase

Table III. Comparison of the effects of P_1 antagonists on responses of the pulmonary and coronary circulation to hypoxia, ATP and adenosine

	Response to hypoxia	ATP	Adenosine
Pulmonary circulation	Constriction	Constriction	Dilatation
P_1 Antagonist	–	–	Block
Coronary circulation	Dilatation	Dilatation	Dilatation
P_1 Antagonist	–	–	Block

and requires the integrity of the purine ring for activation (20). It has also been proposed that there are two ATP binding sites (25).

The distribution of P_1 and P_2 purinoceptors in different tissues has not been fully investigated. Nevertheless, some broad trends seem apparent. For example, P_1 receptors seem predominant in some vascular beds and in the trachea and brain, while P_2 receptors seem predominant in the gastrointestinal tract and urinogenital system. The effects that P_1 and P_2 antagonists have on the responses of various tissues to purinergic nerve stimulation and hypoxia may indicate which purinergic transmitter is involved in these responses (see tables II and III). The primary response of the coronary and pulmonary circulation to hypoxia is vasodilatation and vasoconstriction respectively, and it has been suggested that these responses are mediated by adenosine (3). It has always been anomalous, however, that adenosine does not mimic the vasoconstrictor response that the lung exhibits towards hypoxia and that the vasodilatation produced by hypoxia in the heart is not blocked by theophylline. An alternate explanation is that ATP, rather than adenosine, is the agent producing the responses, since ATP mimics the vasoconstrictor response that the lung exhibits towards hypoxia and since its vasodilator effect on the heart is not blocked by theophylline.

The structure-activity relationships of several analogues of adenine nucleotides and nucleosides have been described (see 10) and the following conclusions drawn as to the requirements of the purine compounds for recognition by the receptor.

The free NH_2 group at the C6 position in the imidazole ring is important for both types of receptor, since alkyl or dialkyl substitutions in the NH_2 group decrease activity to about 5%, and inosine has little pharmacological activity. Also, the ribose moiety is necessary for both types of receptors, since adenine is not active, and the steric configuration of both the purine and ribose rings is essential (8).

The action of adenosine on the methylxanthine-blocked P_1 purinoceptor is extremely structure-dependent. Its agonist activity is enhanced by some substitutions of the purine ring system. For example, the 2-substituted adenosine analogue 2-chloroadenosine is a very effective agonist in brain and coronary vasculature (8). Other analogues with a modified purine ring system (1-methyladenosine and toyocamycin) are more effective in relaxing intestinal smooth muscle than vascular smooth muscle. Substituents that decrease the basicity of the N-1 position or interfere with the hydrogen bonding in the H-6 position decrease the activity in both vascular and intestinal smooth muscle. Agonist action is converted to antagonist action with minor substitutions of the ribose ring system. For example, 2′-, 3′- and 5′-deoxyadenosine are very potent antagonists with no agonist activity in brain, heart, and ileum. *Phillis and Kostopoulos* (24) suggested that the receptor in cerebral cortical neurons exhibits specificity for both the ribose and the base moieties of the molecule. 2′-Deoxyadenosine, however, still relaxed vascular smooth muscle. Hence, it was concluded that intestinal smooth muscle interacts significantly with the C-2 position of the ribose portion of the adenosine molecule, whereas vascular smooth muscle required the adenyl moiety in the molecule for maximal activity. Alkylthio- and alkylamino-substitutions in the adenosine molecule reduce its dilator activity. 2-Methylthio- and 2-methylamino-adenosines are less potent than the corresponding ethyl analogues. A further increase in length of the alkylthio-side-chain enhances dilator activity, but branching of the propyl chain reduces this effect. The duration of coronary dilator activity of these compounds is increased to 5–15 times that of adenosine. Conversion of the adenosine analogues to their monophosphorylated derivatives generally reduces their dilator activity to about one-third.

Structure-activity studies of the P_2 purinoceptor in the intestine have revealed some features that differ from those described in brain and heart (29). For example, compounds with an extended chain of phosphate groupings, or compounds, in which methylene substitution confers resistance of the phosphate chain to enzymatic cleavage, are more potent than ATP in causing relaxation of smooth muscle of intestine (29). Compounds with more than one phosphate grouping cause a rapid relaxation of the taenia coli, except for the alpha-, beta-methylene-substituted nucleotides, which take at least 40% longer to elicit maximal relaxation. Substitution in the 2-position of the purine nucleus by chloro- or methylthio-groups markedly increases the relative activity of the diphosphates and triphosphates.

It has been shown that ultraviolet light between 340 and 380 nm produces responses in guinea-pig taenia coli and rabbit portal vein that mimic precisely the inhibitory response to purinergic nerve stimulation and ATP (14). UV light does not initiate impulses in purinergic nerves since its action is unaffected by tetrodotoxin, nor does it release ATP from nerve terminals. Agents which alter postjunctional responses to ATP and purinergic nerve stimulation also alter the responses to UV light. Since UV irradiation has no action on smooth muscle which is not innervated by inhibitory purinergic nerves, it may be acting on some part of the purinergic receptor complex and may provide a means of investigating the chemistry of purinergic receptors.

Summary

The presence of a non-cholinergic, non-adrenergic component in the vertebrate autonomic nervous system is now well established. Evidence that ATP is the transmitter released from some of these nerves (called "purinergic') includes: (a) synthesis and storage of ATP in nerves: (b) release of ATP from the nerves when they are stimulated; (c) exogenously applied ATP mimicking the action of nerve-released transmitter; (d) the presence of ectoenzymes which inactivate ATP; (e) drugs which produce similar blocking or potentiating effects on the response to exogenously applied ATP and nerve stimulation.

A basis for distinguishing two types of purinergic receptors has been proposed according to four criteria: relative potencies of agonists, competitive antagonists, changes in levels of cAMP and induction of prostaglandin synthesis. Thus P_1 purinoceptors are most sensitive to adenosine, are competitively blocked by methylxanthines and their occupation leads to changes in cAMP accumulation; while P_2 purinoceptors are most sensitive to ATP, are blocked (although not competitively) by quinidine, 2-substituted imidazolines, 2,2'-pyridylisatogen and apamin, and their occupation leads to production of prostaglandin. P_2 purinoceptors mediate responses of smooth muscle to ATP released from purinergic nerves, while P_1 purinoceptors mediate the presynaptic actions of adenosine on adrenergic, cholinergic and purinergic nerve terminals.

References

1 Arunlakshana, O. and Schild, H.O.: Some quantitative uses of drug antagonists. Br. J. Pharmacol. *14:* 48–58 (1959).
2 Berne, R.M.: Cardiac nucleotides in hypoxia: possible role in regulation of coronary blood flow. Am. J. Physiol. *204:* 317–322 (1963).
3 Berne, R.M. and Rubio, R.: Challenges to the adenosine hypothesis for the regulation of coronary flow; in Bloor and Olsson, Advances in experimental medicine and biology; vol. 39, pp. 3–10 (Plenum Press, New York 1973).
4 Burnstock, G.: Purinergic nerves. Pharmac. Rev. *24:* 509–581 (1972).
5 Burnstock, G.: Purinergic transmission; in Iversen, Iversen and Snyder, Handbook of psychopharmacology; vol. 5: Synaptic modulators, pp. 131–194 (Plenum Press, New York 1975).
6 Burnstock, G.: Purine nucleotides and nucleosides as neurotransmitters or neuromodulators in the central nervous system; in Usdin, Hamburg and Barchas, Neuroregulators and psychiatric disorders, pp. 470–477 (Oxford University Press, New York 1977).
7 Burnstock, G.: The purinergic nerve hypothesis; in Purine and pyrimidine metabolism: Ciba Foundation Symposium, No. 48, pp. 295–314 (Elsevier/Excerpta Medica/North Holland, Amsterdam 1977).

8 Burnstock, G.: A basis for distinguishing two types of purinergic receptor; in Bolis and Straub, Cell membrane receptors for drugs and hormones: a multidisciplinary approach, pp. 107–118 (Raven Press, New York 1978).
9 Burnstock, G.: Past and current evidence for the purinergic nerve hypothesis; in Baer and Drummond, Physiological and regulatory functions of adenosine and adenine nucleotides, pp. 3–32 (Raven Press, New York 1979).
10 Burnstock, G.: Cholinergic and purinergic regulation of blood vessels; in Bohr, Somlyo and Sparks, Handbook of physiology, vascular smooth muscle (American Physiology Society, Williams & Wilkins, Baltimore 1980; in press).
11 Burnstock, G.; Cocks, T.; Kasakov, L., and Wong, H.: Direct evidence for ATP release from non-adrenergic, non-cholinergic ('purinergic') nerves in the guinea-pig taenia coli and bladder. Eur. J. Pharmacol. *49:* 145–149 (1978).
12 Burnstock, G.; Cocks, T.; Paddle, B.M., and Staszewska-Barczak, J.: Evidence that prostaglandin is responsible for the 'rebound contraction' following stimulation of non-adrenergic, non-cholinergic ('purinergic') inhibitory nerves. Eur. J. Pharmacol. *31:* 360–362 (1975).
13 Burnstock, G. and Costa, M.: Inhibitory innervation of the gut. Gastroenterology *64:* 141–144 (1973).
14 Burnstock, G. and Wong, H.: Comparison of the effects of UV light and purinergic nerve stimulation on the guinea-pig taenia coli. Br. J. Pharmacol. *62:* 293–302 (1978).
15 Clark, P.B. and Seney, M.N.: Regulation of adenylate-cyclase from cultured human cell lines by adenosine. J. biol. Chem. *251:* 4239–4246 (1976).
16 Coleman, R.A.: Effects of some purine derivatives on guinea-pig trachea and their interaction with drugs that block adenosine uptake. Br. J. Pharmacol. *57:* 51–57 (1976).
17 Drury, A.N. and Szent-Györgyi, A.: The physiological activity of adenine compounds with special reference to their action upon the mammalian heart. J. Physiol., Lond. *68:* 213–237 (1929).
18 Holman, M.E. and Weinrich, J.P.: Effects of calcium and magnesium on inhibitory junctional transmission in smooth-muscle of guinea-pig small intestine. Pflügers Arch. ges. Physiol. *360:* 109–119 (1975).
19 Holton, P.: The liberation of adenosine triphosphate on antidromic stimulation of sensory nerves. J. Physiol., Lond. *145:* 494–504 (1959).
20 Londos, C. and Wolff, J.: Two distinct adenosine-sensitive sites on adenylate cyclase. Proc. natn. Acad. Sci. USA *74:* 5482–5486 (1977).
21 Needleman, P. and Johnson, E.M.: Sulfhydryl reactivity of organic nitrates: tolerance and vasodilatation; in Bevan et al., Proc. Int. Symp. Odense, 1975: Vascular neuroeffector mechanisms; 2nd ed., pp. 208–215 (Karger, Basel 1976).
22 Needleman, P.; Minkes, M.S., and Douglas, J.R.: Stimulation of prostaglandin biosynthesis by adenine nucleotides. Profile of prostaglandin release by perfused organs. Circulation Res. *34:* 455–460 (1974).
23 Olson, L.; Alund, M., and Norberg, K.-A.: Fluorescence microscopical demonstration of a population of gastro-intestinal nerve fibres with a selective affinity for quinacrine. Cell Tissue Res. *171:* 407–423 (1976).
24 Phillis, J.W. and Kostopoulos, G.K.: Adenosine as a putative transmitter in cerebral-cortex studies with potentiators and antagonists. Life Sci. *17:* 1085–1094 (1975).
25 Powers, S.G.; Griffith, O.W., and Meister, A.: Inhibition of carbamyl phosphate synthetase by P1, P5-D1 (Adenosine 5) pentaphosphate – evidence for 2 ATP binding sites (technical note). J. biol. Chem. *252:* 3558–3560 (1977).

26 Rubio, R.; Berne, R.M.; Bockman, E.L., and Curnish, R.R.: Relationship between adenosine concentration and oxygen supply in rat brain. Am. J. Physiol. *228:* 1896–1902 (1975).
27 Satchell, D. and Burnstock, G.: Comparison of the inhibitory effects on guinea-pig taenia coli of adenine nucleotides and adenosine in the presence and absence of dipyridamole. Eur. J. Pharmacol. *32:* 324–328 (1975).
28 Satchell, D.; Burnstock, G., and Dann, P.: Antagonism of the effects of purinergic nerve stimulation and exogenously applied ATP on the guinea-pig taenia coli by 2-substituted imidazolines and related compounds. Eur. J. Pharmacol. *23:* 264–269 (1973).
29 Satchell, D.G. and Maguire, M.H.: Inhibitory effects of adenine nucleotide analogs on the isolated guinea-pig taenia coli. J. Pharmac. exp. Ther. *195:* 540–548 (1975).
30 Turnheim, K.; Pittner, H.; Kolassa, N., and Kraupp, O.: Relaxation of coronary artery strips by adenosine and acidosis. Eur. J. Pharmacol. *41:* 217–220 (1977).
31 Vladimirova, A.I. and Shuba, M.F.: Strychnine, hydrastine and apamin effect on synaptic transmission in smooth muscle cells. Neurophysiology (USSR) *10:* 295–299 (1978).

Prof. G. Burnstock, Department of Anatomy and Embryology,
University College London, London WC1E 6BT (England)

Subject Index